Protective and Functional Coatings for Metallic and Ceramic Substrates

Protective and Functional Coatings for Metallic and Ceramic Substrates

Selected Articles Published by MDPI

MDPI • Basel • Beijing • Wuhan • Barcelona • Belgrade • Manchester • Tokyo • Cluj • Tianjin

This is a reprint of articles published online by the open access publisher MDPI (available at: http://www.mdpi.com). The responsibility for the book's title and preface lies with Cecilia Bartuli, who compiled this selection.

For citation purposes, cite each article independently as indicated on the article page online and as indicated below:

LastName, A.A.; LastName, B.B.; LastName, C.C. Article Title. *Journal Name* **Year**, *Article Number*, Page Range.

ISBN 978-3-03928-448-1 (Pbk)
ISBN 978-3-03928-449-8 (PDF)

Cover image courtesy of the authors (Coatings 2019, 9, 216, doi:10.3390/coatings9040216).

Contents

Preface to "Protective and Functional Coatings for Metallic and Ceramic Substrates"

Coatings here provides a selection of 10 papers, published in 2019, from researchers and institutions based in various countries around the world (nine European, one American, and one Asian), allowing us to appreciate the variety and significance of ongoing research in the wide field of protective and functional coatings.

The most noteworthy investigations conducted in the area of surface protection are currently proceeding with an identically fast pace in the twofold direction of a deeper and increasingly reliable knowledge of degradation and protection mechanisms, and of the technological optimization of new materials selection and design, coatings deposition processes, and characterization methods.

A representative essay of the worldwide efforts toward more durable surfaces can be read in this short collection. Both organic and inorganic coatings are included among the protection strategies, from a large variety of deposition processes, with interesting examples of organic-inorganic composites being proposed (such as PVC-ZnO nanocoposites or MoS2-sunflower oil combinations). Major attention is devoted to protection from both electrochemical and chemical corrosion of different metallic alloys (stainless steel, bronze, and aluminum) and to advanced SiC-SiC composites exposed to the aggressive environments characteristic of gas turbines.

The delicate issues related to the modeling and experimental evaluation of mechanical properties of coatings are comprehensively investigated in papers concerning cavitation erosion and sliding wear resistance and stress-dependent elasticity of Ti–Al–N-based thin films.

A dedicated space was purposely devoted in the papers selection for examples of applicability of an advanced characterization technique (attenuated total reflectance Fourier transform infrared microscopy) for the evaluation of bituminous coatings from reclaimed asphalt, as representative of the important ongoing efforts to achieve reliable and affordable characterization and classification (and standardization) procedures of both newly conceived or combined materials.

Finally, as an example of the wide field of research on innovative functional coatings, a study is presented of highly active Mo-based alloys deposited in the form of films as electrocatalytic material for hydrogen evolution reaction.

A comprehensive review of directed energy deposition additive manufacturing of metallic components completes the collection, highlighting the complex physico-chemical interactions between matter and heat, which are at the basis of Additive Manufacturing (AM) production processes, and whose knowledge and modeling capacity owes much to the experience acquired in the study of high temperature coating deposition processes.

—Prof. Dr. Cecilia Bartuli
University of Rome "La Sapienza"

coatings

MDPI

Article

Effect of Synthesis Conditions on the Controlled Growth of MgAl–LDH Corrosion Resistance Film: Structure and Corrosion Resistance Properties

Muhammad Ahsan Iqbal * and Michele Fedel *

Department of Industrial Engineering, University of Trento, via Sommarive 9, 38123 Trento, Italy
* Correspondence: muhammadahsan.iqbal@unitn.it (M.A.I.); michele.fedel@unitn.it (M.F.); Tel.: +39-320-022-7519 (M.A.I.); +39-461-285-354 (M.F.)

Received: 10 December 2018; Accepted: 2 January 2019; Published: 7 January 2019

Abstract: In this study, a series of MgAl–layered double hydroxide (LDH) thin films were synthesized by a single step hydrothermal process at different synthetic conditions on AA6082, and the combined effect of reaction temperatures and crystallization times on in situ growth MgAl–LDH structural geometry, growth rate, and more importantly on the corresponding corrosive resistance properties are briefly discussed. The synthesis of LDH was performed at reaction temperatures of 40, 60, 80, and 100 °C, while the treatment time was varied at 12, 18, and 24 h. The as-prepared synthetic coatings were characterized by scanning electron microscopy (SEM), energy dispersion spectroscopy (EDS), X-ray diffraction (XRD) and Fourier transform infrared spectroscopy (FT-IR), while the corresponding corrosion protection efficiency of the developed coating was studied through potentiodynamic polarization studies and electrochemical impedance spectra. The findings demonstrated that extended crystallization time and reaction temperature impart a significant effect on the oriented growth of layered double hydroxide, the surface morphology, and on the film thickness, which had a remarkable influence on the LDH corrosion resistance ability. The LDH coated specimen developed at 100 °C for 18 h reaction time showed a more compact and dense structure compared to the traditional platelet structure obtained at 80 °C for 24 h crystallization time, and interestingly that compact structure exhibited the lowest corrosion current density, up to five orders of magnitude lower than that of bare AA6082.

Keywords: MgAl–layered double hydroxide; EIS; XRD; FT-IR; AA6082; corrosion protection

1. Introduction

Aluminum and its alloys are considered promising materials for a number of applications due to their high strength to weight ratio and relatively high thermal and electrical conductivities, along with their abundance and low price. In addition, aluminum forms a passive oxide layer on its surface, which increases its corrosion resistance. On exposure to an acidic or alkaline medium, especially chloride-incorporating media, the protective oxide layer is damaged and aluminum becomes susceptive to corrosion. Various series of aluminum alloys have been developed to improve the properties of aluminum with respect to their usage applications. AA6xxx is relatively a new class of aluminum alloys which contain mainly additions of manganese, magnesium, silicon, and a small amount of copper [1–4]. Although these alloys show relatively high general corrosion resistance, they are susceptible to forms of localized corrosion, especially pitting and intergranular corrosion [5,6]. Therefore it has become indispensable to develop an efficient, environmentally friendly protective coating on aluminum and its alloys.

Chromate-containing conversion coatings were employed for a long time to protect aluminum alloys but since those coatings contain hexavalent chromate compounds, which have been proved toxic for the environment and hazardous to health [7,8], efforts have been made to develop suitable alternatives. A number of approaches have been reported, including conversion coatings [9,10], magnetron sputtering [11,12] anodizing [13,14], sol-gel synthesis [15,16], and self-assembly [17,18], and polymer coatings [19,20] to develop non-chromate based corrosion resistance coatings. Layered double hydroxide, a promising type of chemical conversion coatings and also known as hydrotalcite-like compounds or anionic clays, have had prominent attention in the fields of biomedical science, applied chemistry, and environmental purifications and recently as corrosion resistant coatings for metals [21,22]. Generally, two different methods have been employed to fabricate layered double hydroxide, the single step in situ approach [23,24] and the other a colloidal assembly technique [25,26]. Coatings fabricated using the two-step method improved the corrosion resistance of their substrates; however, the poor adhesion of the film to the substrate was reported and also the fabrication process itself appears slightly complicated [27]. Numerous works have been reported to fabricate different types of layered double hydroxide to protect the light metals alloys [27–30]. The growth rate of crystals and their size and distribution can be controlled by adjusting the crystallization time and the reaction temperature. The properties and structural characteristics of MgAl–layered double hydroxide (LDH) actively depend upon the fabrication method, operating parameters, and conditions used for the fabrication [31–34]. Therefore, the optimization of the synthesis parameters plays an important role in developing a suitable structure for numerous applications. In our previous work, we reported the effect of different salt concentrations on LDH structural growth to obtain various distinct LDH morphologies as well as their effect on corrosion resistance properties [35]. However, the combined effect of extended reaction temperature and the aging time on in situ growth MgAl–LDH structural growth rate without using any complexation agents (used to promote specific LDH structural growth) to further their impact on their corresponding corrosive resistance behavior has not yet been thoroughly investigated. In this work we succeeded to synthesize a series of MgAl–NO_3 LDH film on the surface of AA6082 by using magnesium salt only, and developed a range of balanced combinations of reaction temperature and aging time, at constant initial cationic concentration (Mg^{2+}) to investigate in detail the impact of the above-mentioned parameters on LDH geometry, structural growth, morphology, and their effect on the corresponding corrosion resistance properties. In particular, in this study, the one-step in situ growth method was used to develop MgAl–LDH coatings on AA6082 at different combinations of extended reaction temperature and crystallization time without using any surfactants or complexation reagents. This was done in order to understand the effect of the mentioned parameters on the geometry of LDH crystallites, film growth, and on the deposition rate to explain the relationship of LDH structural variations with its anticorrosion behavior. This work provides insight into the corrosion resistance properties of MgAl–LDH, and into the correlation between the electrochemical response of the coatings and their structural properties.

2. Experimental

2.1. Materials

The AA6082 extruded bar was purchased from Metal Center (Trento, Italy), which mainly consists of (0.60%–1.20%) magnesium, (0.50) iron, (0.70%–1.30%) silicon, (0.10%) copper, (0.40%–1.00%) manganese and of a balance percentage of aluminum. The $Mg(NO_3)_2 \cdot 6H_2O$, (purity 98%), NH_4NO_3 (purity 95%)and NaOH (purity $\geq$98%) were purchased from Sigma-Aldrich Corporation (Saint Louis, MO, USA).

2.2. Synthesis of Mg–Al Layered Double Hydroxide Film

The AA608 specimens were initially ground with silicon carbon paper, starting from 500 grit paper to 1000, 2400, and 4000-grit respectively. The samples were cleaned with deionized water and further ultrasonically in pure ethanol for 15 min. Lastly, the specimens were immersed in a 0.1 M aqueous NaOH solution for one minute to etch the oxide layer on the surface of the alloy.

A 0.05 M ($Mg(NO_3)_2{\cdot}6H_2O$) solution was prepared in 1000 mL deionized water and the pH of the solution was adjusted to 10.0 by the dropwise addition of ammonium hydroxide solution. The solution was poured into a four bottleneck heating flask (Sigma-Aldrich Glassware) and pretreated aluminum samples (3.14 cm^2) were immersed vertically in the solution. The experiments were performed at 40, 60, 80, and 100 °C reaction temperature, for 12, 18, and 24 h, under nitrogen bubbling. The samples were washed with deionized water after the experiments completion and were dried at room temperature. Figure 1 represents the general schematic diagram of the layered double hydroxide.

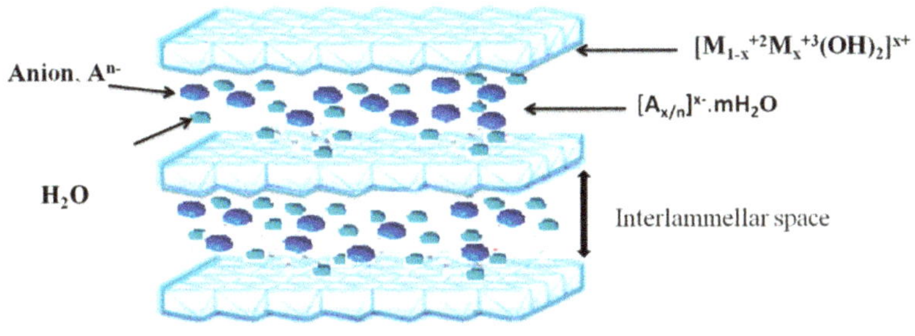

Figure 1. Schematic illustration of layered double hydroxides.

3. Characterization

A scanning electron microscopy (JEOL-IT300 microscope equipped with an EDS detector, JEOL Ltd, Tokyo, Japan) was used to analyze the morphology and the microstructural features of MgAl-LDH, while the elemental analysis of the thin film was studied through energy dispersive electron microscopy (SEM-EDX) in planner mode. XRD patterns of the calcined and uncalcined LDH coated samples were recorded by X-ray diffraction (XRD) (X'Pert High Score diffractometer, Rigaku, Tokyo, Japan) at ambient conditions by using cobalt $K\alpha$ (λ = 1.789 Å) emission source at 10 mA and 30 kV conditions. The step size of 0.005° was adjusted in the 2θ range of 5°–110°. The peak reflections at the (003) planes of layered double hydroxide were studied to measure the basal spacing and to define the full-width half maximum (FWHM) for the measurement of crystallite size (D), by using the Scherrer formula. The lattice parameters of Mg–Al layer double hydroxide were calculated by using Equation (1) [34], where d is the lattice spacing calculated by Braggs law.

$$\frac{1}{d^2} = \frac{4}{3}\left(h^2 + hk + \frac{k^2}{a^2}\right) + \frac{l^2}{c^2} \quad (1)$$

Fourier transformed infrared spectra (FTIR), Varian 4100 FTIR Excalibur Series instrument (Varian, Santa Clara, CA, USA), in the attenuated total reflectance (ATR) mode were recorded to analyze surface functional groups and the chemical bonding of the samples, in the range of 550–4000 cm^{-1} with a 4 cm^{-1} resolution and at 32 scans, by using a diamond crystal as Internal Reflective element (IRE). The electrochemical measurements were conducted (PAR Parstat 2273, Ametek, Berwyn, PA, USA) 0.1 M NaCl solution to understand the anti-corrosion properties performed at ambient conditions under Faraday cage to minimize the external noise on the system. A relatively diluted electrolyte (0.1 M) was employed in order to better highlight the differences between the investigated samples. A traditional three electrode setup was used, where the platinum mesh was used as a counter electrode while MgAl–LDH and Ag/AgCl/KCl (+210 mV vs. Standard Hydrogen Electrode (SHE)) were used as working and reference electrode respectively. The MgAl–LDH was exposed to the electrochemical solution for 30 min in 0.1 M NaCl solution before the measurement, for the system stabilization. The polarization curves were obtained with a sweep rate of 2 mV/s. The impedance measurements (EIS) were performed at the open circuit potential (OCP) over frequencies ranging from 100 kHz down to 10 mHz at a 5 mV (rms) amplitude.

4. Results and Discussion

4.1. XRD Analysis

Figure 2a,b shows the XRD pattern of MgAl–LDH thin film developed on the AA6082 at various temperature and crystallization time periods. The XRD pattern of the MgAl layer double hydroxide at 40 °C–24 h and at 60, 80, 100 °C for 12, 18, and 24 h crystallization time, demonstrated the characteristic peaks of layered double hydroxide structure, described in the literature [35], further the sharpness of the peaks suggested an ordered structure of LDH layers. It can be seen that synthesis temperature and crystallization time have an influential effect on characteristic peaks. The (003), (006), (009), (110), and (113) characteristics peaks of crystal planes are related to the MgAl–NO_3 LDH film. The intensity and broadness of the reflection peaks vary with the increase of temperature and t also depend upon crystallization aging time. When the samples were synthesized at 60 °C and lower aging time (12 and 18 h), the pattern exhibited low intensity and a broader peak which describe their low crystallinity. At higher temperature range, for instance, 80 and 100 °C, the reflection peaks exhibited sharp and high intense narrow peaks which describe well the crystalline and ordered structure. The intense reflection peaks of (003) at low 2θ value exhibited an interlayer distance of ≈0.80 nm and the reflection peak could further be used to calculate the cell parameter "*c*" by the correlation "$c = 3d_{003}$", while the d_{110} interlayer thickness was used to calculate the cell parameter "*a*" by the correlation "$a = 2d_{110}$". The basal spacing of MgAl LDH slightly shifted to lower angle with the effect of enhanced temperature and crystallization aging time, indicating the strong intercalation of NO_3^- ions. However due to the formation of carbonate anions, small absorption peaks of d_{003} at 40 °C–24h, 60 °C–18 h were also observed. Fourier transform-infrared spectroscopy (FT-IR) spectra further confirmed the presence of CO_3^{2-} nions. The interlayer thickness, lattice constants of "*a*" and "*c*", and crystallite size for Mg–Al LDHs thin film on AA6082 are listed in Table 1.

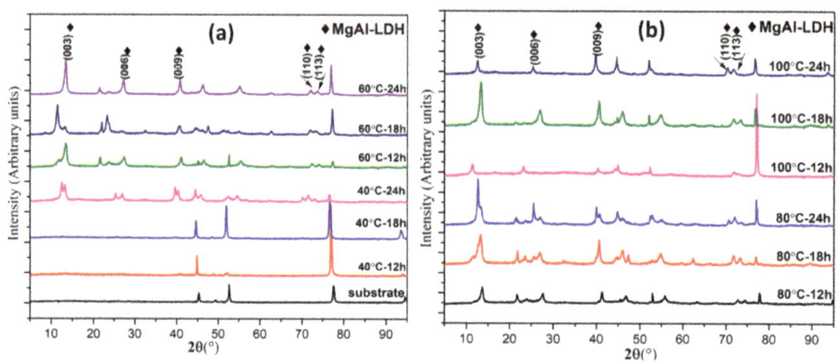

Figure 2. The XRD spectra of Mg-Al layered double hydroxide (LDH) film samples developed on aluminum alloy AA6082 obtained at various reaction temperature and crystallization times.

4.2. Fourier Transform Infrared Spectroscopy (FTIR)

The selective samples, synthesized at various temperatures and crystallization times, were further investigated by FTIR analysis. FTIR spectra in attenuated total reflection mode were recorded, as shown in Figure 3. All the samples showed almost similar types of features. The broadbands displayed in the range of 3370–3427 cm^{-1} were assigned to OH group stretching and the absorption band around, 1627–1633 cm^{-1} was caused due to the flexural oscillation peaks of interlayer water molecules [36]. Moreover, the absorption peaks around 1350 cm^{-1} were assigned to the asymmetric stretching bond of intercalated NO_3^- [37]. The small peak around 1520 cm^{-1} corresponded to the CO_3^{2-} ions, which shows the presence of a small number of carbonate ions in the LDH phase. The bond at 655, 751, and

1202 cm^{-1} may be associated with Al–OH stretching [38]. The absorption peaks around 550–770 cm^{-1} correspond to the lattice vibration of metal–oxygen bonds (M–O) [39].

Table 1. Cell parameters and sizes of the coherent domains determined for the Mg−Al layered double hydroxide (LDH) precursor powders from X-ray diffraction.

Sample	Cell Parameter, *a* (nm)	Cell Parameter, *c* (nm)	Interlayer Distance, d_{003} (nm)	Interlayer Distance, d_{110} (nm)	Crystallite Size, *D* (nm)
40 °C–24 h	0.308	2.329	0.776	0.154	11.214
60 °C–12 h	0.306	2.296	0.765	0.153	12.908
60 °C–18 h	0.305	2.668	0.889	0.152	7.608
60 °C–24 h	0.305	2.294	0.765	0.152	11.657
80 °C–12 h	0.303	2.275	0.759	0.151	6.685
80 °C–18 h	0.306	2.461	0.820	0.153	6.987
80 °C–24 h	0.311	2.412	0.804	0.155	8.465
100 °C–12 h	0.311	2.691	0.897	0.155	5.929
100 °C–18 h	0.311	2.391	0.797	0.156	6.165
100 °C–24 h	0.311	2.463	0.821	0.157	6.411

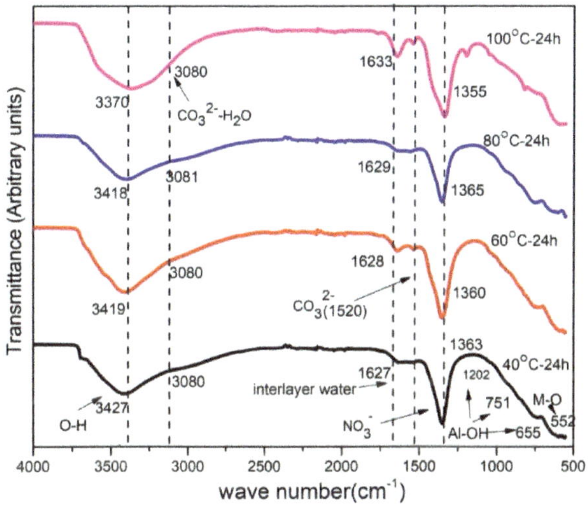

Figure 3. ATR-FTIR spectra of Mg–Al LDH powder, scraped from as-prepared MgAl–LDH coated samples fabricated at different reaction temperatures.

4.3. Scanning Electron Microscopy (SEM)

As revealed from the SEM observations, shown in Figure 4, a compact and uniform interviewed LDH structure was obtained by the increase of reaction temperature and of aging time, moreover distinct surface morphological variations were also observed. Initially, the LDH granules started to originate and developed into distinct platelet structure with increase of reaction temperature and crystallization time, while on further increase of the temperature and aging time, the nano-sheets of the LDH structure fused to form a compact blade like LDH structure. For example at 80 °C temperature and 24 h aging time, a fully groomed curved shaped platelet morphology was obtained which fused on further increase in temperature and appeared as a compact needle-shaped structure. The atomic composition of LDH, calculated by energy disperse spectroscopy (EDS) plane scanning, is listed in Table 2. The EDS findings showed that the coated structure is mainly composed of Mg, Al, N, and oxygen. A small amount of carbon is also evident in the EDS study, caused by contamination from air, which results in the formation of carbonate anions. The atomic ratio of Mg/Al varies with the effect of temperature and crystallization time. Table 2 showed that with a higher reaction temperature and crystallization time, a higher ratio of Mg/Al is obtained. The oxygen content also varied with the influence of temperature and time. The atomic ratio of Mg:Al lies in the range of 2.5:1–3.87:1, which provides a rough approximation of the MgAl LDH assembly. Note that the Mg:Al ratio increases with the duration of the treatment regardless of the temperature. The duration of the treatment seems therefore to affect the chemical composition of the LDHs which enrich with Mg during the long immersion time. The possible formation of MgAl LDH on the AA6082 surface may start with the dissolution of aluminum in the basic solution to release Al^{3+} ions (aluminum surface is partially dissolved on contact with the high pH solution and generated aluminum oxides) [40], which react with the water to form $Al(OH)_3$ while the final step is related to the precipitation of Mg^{2+} and OH^- on the surface of the $Al(OH)_3$ surface to form the pre-MgAl LDH hydroxide mixture. The ammonium nitrate continuously hydrolyzed to form NO_3^- in the system, along with the release of ammonium ions which led to progressively keep the pH in the range of 10. Finally, the divalent Mg^{2+} ions in $Mg(OH)_2$ were substituted by the trivalent Al^{3+} ions, which result in the coexistence of $Al(OH)_3$ and $Mg(OH)_2$ to form a stable hydrotalcite-like LDH structure. On the other hand, the hydroxyl ions present inside the galleries were exchanged by NO_3^- ions present in the solution and as the average atomic ratio of Mg to Al is approximately around three, as shown in Table 2 (EDS analysis), it can be concluded that the final film of $Mg_3Al(OH)_2{\cdot}xNO_3{\cdot}yH_2O$ was formed on the substrate of AA6082 and formed stacking of LDH layers one above another to form the LBL configuration (Figure 1). The obtained EDS findings are in agreement with the XRD results.

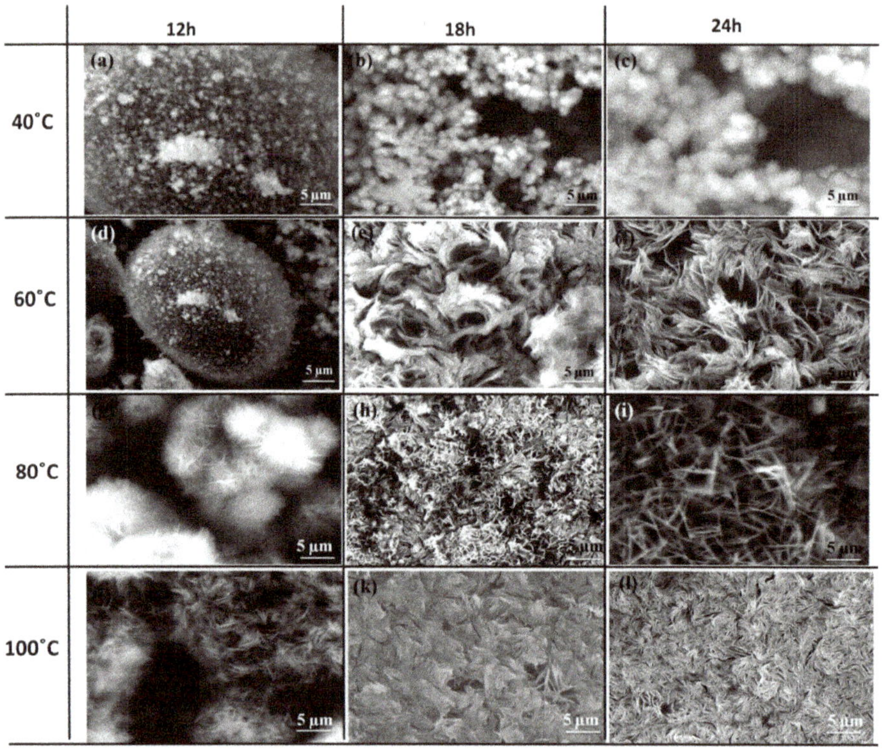

Figure 4. SEM images of LDH films developed at various reaction temperatures and crystallization times.

Table 2. EDS study of atomic percentage at various reaction temperature and crystallization times.

Sample	Mg (at.%)	Al (at.%)	N (at.%)	O (at.%)	Mg/Al
40 °C–24 h	17.6	6.8	3.1	64.7	2.6
60 °C–12 h	18.1	7.2	3.0	60.7	2.5
60 °C–18 h	26.2	7.5	3.3	57.4	3.5
60 °C–24 h	27.5	7.5	4.2	57.0	3.7
80 °C–12 h	19.3	7.3	3.1	58.7	2.6
80 °C–18 h	27.2	7.6	4.1	53.8	3.6
80 °C–24 h	24.7	6.6	4.7	56.2	3.8
100°C–12 h	24.5	7.6	4.2	53.9	3.2
100 °C–18 h	28.6	7.5	4.7	53.9	3.8
100 °C–24 h	28.7	7.4	4.3	63.1	3.9

The results from SEM (Figure 4) suggest that the synthesis conditions promote the formation of uniform and quite ordered structures. As far as the appearance of the coatings is concerned, it seems that on increasing the treatment temperature from 60 to 100 °C, the LDHs layer becomes more and more compact and homogenous. Also the duration of the treatment plays a similar role: prolonged immersion in the solution promotes the formation of homogeneous LDHs structures in which the platelets are closely packed.

The effect on the film thickness of the synthesis parameters is reported in Figure 5, which shows also the cross-sectional image of MgAl LDH at 80 °C and 18 h (reported as an example). The results of the film thickness indicated that with the increase of reaction temperature and treatment time, the film thickness of coating increased. The film thickness increased from about 20.8–73.5 μm with the increase of the processing parameters. The high temperature and aging time in this study were shown to promote the growth rate of LDH formation, confirmed also by XRD and SEM analysis. It was observed that a rapid increase in film thickness was found for 100 °C temperature and 24 h treatment time. However, with the increase of thickness, the interface of substrate and coating become blurred and also caused an increase in defect density, which promoted loose interaction with the substrate. In fact, some pores were observed at the interface of coating and substrate with the increase of coating thickness. This fact is expected to negatively influence the corrosion protection ability of LDH.

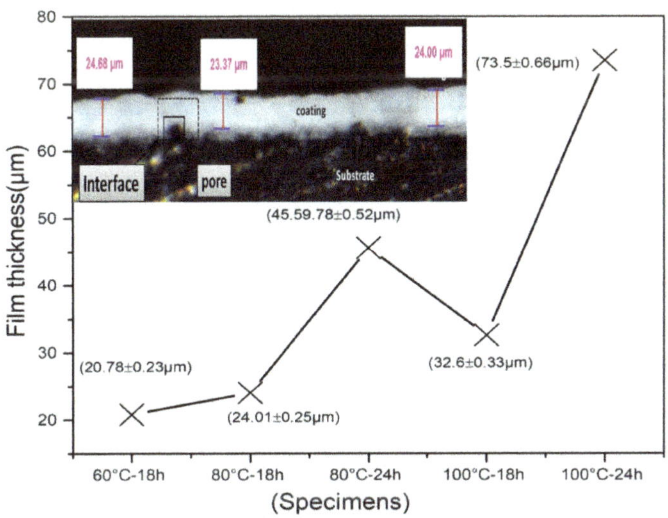

Figure 5. MgAl LDH thickness of the film coated AA6082 estimated from the cross-sectional optical microscopic images as a function of reaction temperature and treatment time, (for representation, insight is the optical images of LDH at 80 °C and 18 h reaction conditions.).

4.4. Electrochemical Study

The as-prepared MgAl–LDH anticorrosion behavior was investigated through electrochemical studies. Figure 6 exhibits the potentiodynamic polarization curves of all the tested samples, while the corrosion current density and corrosion potential are listed in Table 3. With the increase of synthetic reaction temperature and crystallization time, the corrosion current density decreases, and also a significant increase in the film is observed. The reduction of both the anodic and cathodic currents suggests an increase in corrosion resistance. The polarization curves of MgAl LDH coating obtained from the specimens developed at 60 °C–24 h, 80 °C–18 h, 80 °C–24 h, 100 °C–18 h and 100 °C–24 h showed a noticeable decrease in corrosion of both anodic and cathodic current density compared to the bare AA6082. The significant reduction of the current densities suggests that the MgAl LDH thin films can help to decrease the corrosion rate of the substrate. All the developed films on AA6082 showed lower corrosion current density along with a shift of the corrosion potential to higher values compared to bare AA6082. As one can see, for almost all the investigated samples a rise in the corrosion potential was observed as the crystallization time increased. The MgAl LDH is believed to provide corrosion protection due to the following two mechanisms: (1) the barrier effect, as they are dielectric materials which protect the metal surface by avoiding interaction with the metal substrate; (2) by entrapping Cl^- ions and releasing nitrates [41]. In the case of the polarization curves, it is likely that the main protection mechanism observed is the physical barrier effect provided by the LDH layer which promotes a decrease in the flow of current.

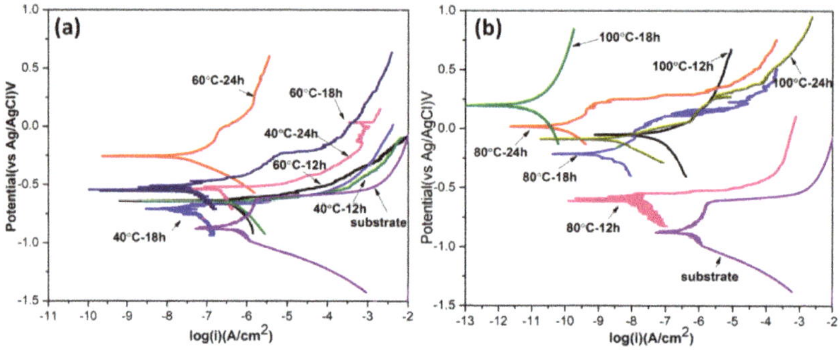

Figure 6. Polarization curves of bare AA6082 and the coated AA6082 in 0.1 M NaCl aqueous solution.

Table 3. The electrochemical parameters estimated from the polarization data in Figure 6.

Specimens	*E* (vs. Ag/AgCl) (V)	*I* (μA/cm²)
AA6082	−0.879	0.46556
40 °C–12 h	−0.679	0.33138
40 °C–18 h	−0.721	0.07629
40 °C–24 h	−0.564	0.28947
60 °C–12 h	−0.683	0.40345
60 °C–18 h	−0.519	0.07524
60 °C–24 h	−0.254	0.05214
80 °C–12 h	−0.623	0.03121
80 °C–18 h	−0.243	0.00156
80 °C–24 h	+0.016	0.00024
100 °C–12 h	−0.099	0.03903
100 °C–18 h	+0.241	0.00001
100 °C–24 h	−0.122	0.00263

In order to further understand the corrosion resistance ability of MgAl LDH, EIS analysis was performed. Figures 7 and 8 show the impedance modulus and phase plots, respectively. At low frequency (about 0.01 Hz), higher impedance modulus values ($|Z|_{0.01}$) are a rough estimation of the corrosion resistance properties [42]: It can be seen that the samples 100 °C–24 h and 100 °C–18 h exhibit the highest value of $|Z|_{0.01}$ among the investigated samples. Furthermore, it is clear that $|Z|$ values follow a general tendency of corrosion resistance, higher temperature, and crystallization time and exhibit greater propensity of corrosion resistance, according to the higher thickness of the LDH layers. These results are consistent with the results obtained from the potentiodynamic curves. The specimens fabricated at 100 °C–24 h showed comparatively a lower corrosion resistance than 100 °C–18 h, which may be due to their greater film thickness which causes more defects in the coating itself and/or pores at the interface with the substrate. The impedance results suggest that the high temperature and aging time lead to an increase in corrosion resistance, which is probably due to increased crystallization of the layered double hydroxide and to a thickening of the layer as well. When the temperature conditions and aging time are moderate, the enhancement in terms of corrosion resistance is less pronounced. In addition, the $|Z|$ values of the MgAl LDH film developed at different reaction conditions follow a general tendency: the higher the hydrothermal reaction temperature, the higher is the corresponding $|Z|$ value obtained. The thickness of the film also plays a vital role in corrosion protection, and $|Z|$ values vary with the increment in film thickness. The sample (100 °C–18 h) showed around 33 μm film thickness and also the highest corrosion resistance among all the processed samples.

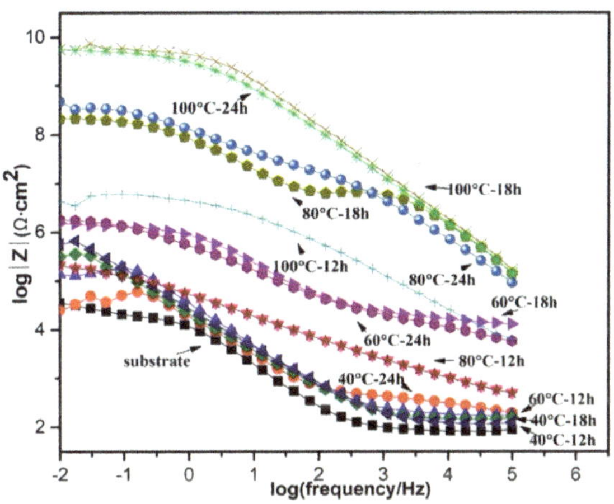

Figure 7. EIS spectra (Bode plots of log($|Z|$) vs. log(frequency)) of bare AA6082 alloy and AA6082 alloy coated with LDH film at various reaction temperatures and crystallization times in 0.1 M NaCl solution.

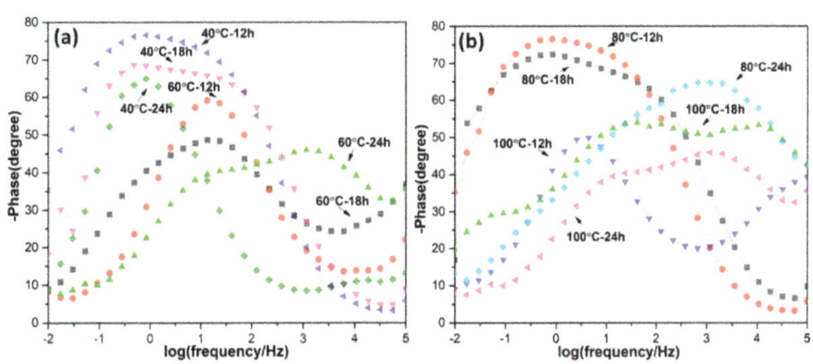

Figure 8. EIS spectra (Phase angle vs. log(frequency)) of bare AA6082 alloy and AA6082 alloy coated with LDH film at various reaction temperatures and crystallization times in 0.1 M NaCl solution.

Figure 9 shows the evolution with an immersion time of the impedance modulus of the most representative samples among those investigated (data collected after 1, 3 and 7 days of immersion are reported). All the samples show a quite relevant decay of the protection properties during the 7 days of continuous immersion in the electrolyte. Regardless of the synthesis conditions, after one week of immersion, the low-frequency impedance is in the 10^5–10^6 Ω cm^2 range for all the investigated samples. This value is in any case remarkably higher than the $|Z|_{0.01}$ value of the bare substrate at the beginning of the immersion in the electrolyte ($|Z|_{0.01} \approx 10^4$ Ω cm^2). The decrease in impedance observed for the investigated coatings is likely to rely on the presence of porosity and defect in the coating and at the metal/LDH layer interface. According to optical microscope observations of the LDH layer and LDH/metal interface, thicker coatings (i.e., with synthesis conditions which promote higher LDH growth rate) are more defective and porous. This would explain the evolution of the impedance spectra with time: at the beginning, the LDH layers act as a barrier against electrolyte permeation; as time elapses, the solution moves through the defects and pores thus reaching the metal/coating interface. The latter process promotes a decrease in impedance over all the frequency range since coating resistance is reduced and the Faradic process is promoted.

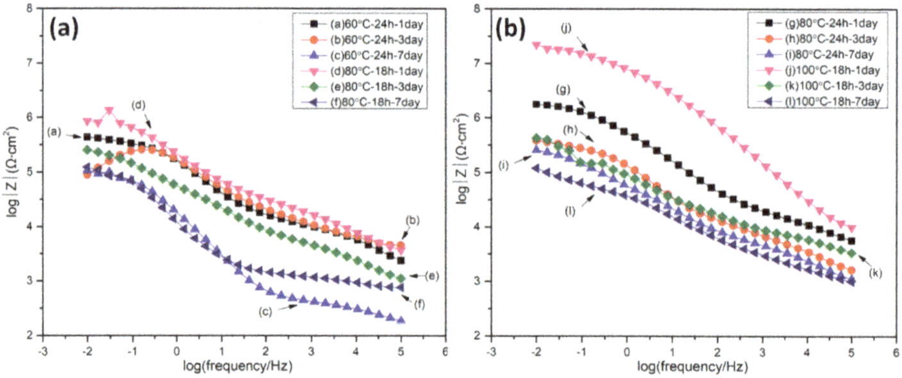

Figure 9. Bode plots of AA6082 samples, coated with MgAl–NO_3 LDH thin film after immersion in 0.1 M NaCl: (**a**) 60 °C–24 h and 80 °C–18 h sample from 1 to 7 days immersion; (**b**) 80 °C–24 h and 100 °C–18 h 1 to 7 days immersion time.

The EIS spectra reported in Figure 9 were fitted in order to get some insight into the effect of the different parameters on the corrosion protection properties of the investigated coatings. According to Figure 8, all the samples show two relaxation processes, which are located in the middle–high or middle–low frequency range. Notice that the thickness, as well as the defectiveness, of the coatings strongly affect the electrochemical response of the diverse samples. For this reason, it is not surprising to observe dispersion of the relaxation processes in the investigated frequency range. According to the literature [43,44], the EIS response of the LDH covered samples can be analyzed employing a $R_s(CPE_{LDH}(R_{LDH}(CPE_{dl}R_{ct})))$ electrical equivalent circuit. In this circuit, R_s stands for the resistance of the electrolyte. The relaxation process at the higher frequencies is fitted employing a resistance, R_{LDH}, which indicates the pore resistance of the LDH layer and a constant phase element, CPE_{LDH}, which represents the dielectric properties of the layered double hydroxide coatings. The relaxation process at the lower frequencies is fitted employing a resistance, R_{ct}, which is related to the superimposition of the contribution of the charge-transfer process and of the aluminum oxide and a constant phase element, CPE_{dl}, which represents the double layer capacitance contribution, overlapped to that of the dielectric properties of the aluminum oxide. According to the mathematical representation of a CPE, (i.e. $Z_{CPE} = 1/(Q(\omega j)^{\alpha})$ the parameters Q and α were employed to describe the dielectric response of the electrodes. Table 4 shows the fitting results for the investigated samples. Notice that the parameters

α_{LDH} and α_{dl} are almost always far from unity. For this reason, it is not possible to attribute any precise physical meaning to the pre-exponential factors Q_{LDH} and Q_{dl}. As far as the R_{LDH} and R_{ct} parameters are concerned, one can observe a certain scattering during immersion time for the samples 60 °C–24 h and 80 °C–18 h, probably due to the porosity of the coatings and to the consequent localization of the corrosion attack promoted by chlorides. On the other hand, the samples 80 °C–24 h and 100 °C–18 h show a smoother decrease in the corrosion protection properties (described by R_{LDH} and R_{ct}) which indicates that the coatings are more compact (as suggested by the relatively high values of R_{LDH}) and protective (as suggested by the relatively high values of R_{ct}).

Table 4. Evolution with time of the fitting parameters R_{LDH}, Q_{LDH}, α_{LDH}, R_{ct}, Q_{dl}, and α_{dl}.

Sample	Immersion Time	R_{LDH} (kΩ cm²)	Q_{LDH} (Ω^{-1} cm^{-2} s$^{\alpha}$)	α_{LDH}	R_{ct} (kΩ cm²)	Q_{dl} (Ω^{-1} cm^{-2} s$^{\alpha}$)	α_{dl}
60 °C–24 h	1 day	13.0	4.2×10^{-7}	0.51	420.0	1.4×10^{-6}	0.72
	3 days	96.30	1.7×10^{-6}	0.31	828.7	1.1×10^{-6}	0.71
	7 days	98.3	1.1×10^{-5}	0.85	0.5	7.8×10^{-6}	0.46
80 °C–18 h	1 day	5705.0	5.1×10^{-6}	0.91	0.9	7.2×10^{-5}	0.46
	3 days	50.1	7.2×10^{-6}	0.51	270.2	9.9×10^{-6}	0.65
	7 days	118.1	1.7×10^{-5}	0.85	2.2	7.9×10^{-5}	0.19
80 °C–24 h	1 day	41.0	4.6×10^{-7}	0.54	2095.0	8.1×10^{-8}	0.77
	3 days	11.2	2.0×10^{-6}	0.43	372.9	1.8×10^{-6}	0.69
	7 days	7.8	2.5×10^{-5}	0.71	354.8	1.0×10^{-5}	0.93
100 °C–18 h	1 day	6135.0	2.0×10^{-8}	0.69	18420	7.8×10^{-8}	0.68
	3 days	13.0	4.2×10^{-7}	0.51	418.4	1.4×10^{-6}	0.72
	7 days	104.2	8.8×10^{-6}	0.46	97.4	1.2×10^{-4}	1.0

To better understand as to what extent the investigated LDH layers provide corrosion protection, the numerical values of R_{LDH} and R_{ct} obtained by fitting the experimental curves were compared with data from the literature. In Table 5, a comparison of the fitting parameters R_{LDH} and R_{ct} after about 1 day (or less) of immersion in a sodium chloride electrolyte is reported. The value of R_{LDH} for the coatings developed in this study is remarkably higher compared to the data reported in the literature. However, one should consider that: (1) in this study the crystallization treatment was prolonged in order to obtain relatively thick coatings, while in the literature very often only a thin conversion layer of LDHs was investigated; (2) the electrolyte employed in this study is more dilute than 3.5 wt % NaCl (0.1 M ≈ 0.58 wt %): for this reason, higher resistance values are expected. Also the parameter R_{ct} for the samples investigated in this study is one of the highest compared to other works reported in the literature (except the LDH reported in [43], which contains vanadates based on corrosion inhibitors). However, also in this case it has to be considered that the electrolyte employed is more dilute compared to the NaCl concentration employed in the studies reported in [44–46].

Table 5. Comparison with literature data referring to LDH conversion layers/coatings on aluminum alloys (n.p.: not provided).

LDH	NaCl Concen.	Time	R_{LDH} (kΩ cm²)	R_{ct} (kΩ cm²)	Ref.
Li/Al	3.5 wt %	0 h	2.2	6.49×10^3	[43]
Mg/Al	3.5 wt %	1 day	n.p.	5.88	[44]
Li/Al	3.5 wt %	1 day	0.8	0.18×10^3	[45]
Zn/Al (+VO_x)	0.05 M	1 day	18.2	7.96×10^8	[46]
Mg/Al 80 °C–24 h	0.1 M	1 day	41.0	2.09×10^3	This work
Mg/Al 100 °C–18 h			6135.0	18.4×10^3	

The EIS results seem also to be in accordance with the visual observation of the immersed samples as shown in Figure 10: optical microscope photos were taken after the 7 days of immersion in 0.1 M NaCl solution. The bare aluminum sample was badly corroded after the immersion and was covered with a thick layer of corrosion products. Notice that cracks and blister are observed on the LDH layers, in accordance with the remarkable decrease in R_{LDH} and R_{ct}.

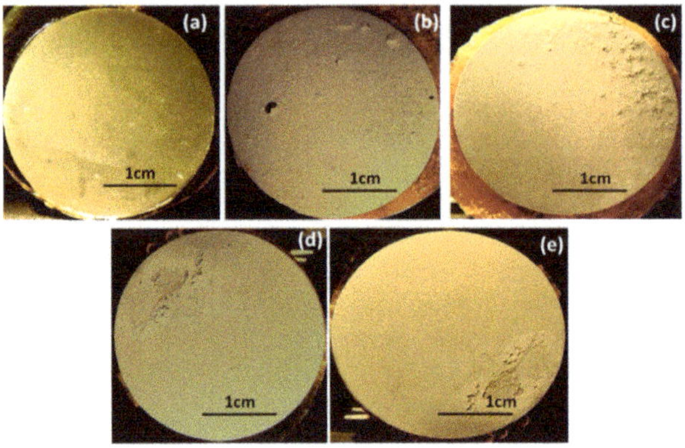

Figure 10. The specimens after 7 days of immersion in a solution of 0.1 M NaCl: (**a**) bare aluminum surface; (**b**) coated sample at 60 °C–24 h; (**c**) 80 °C–18 h; (**d**) 80 °C–24 h; and (**e**) 100 °C–18 h.

The surface morphologies of the MgAl–LDH after 7 days immersion in 0.1 M NaCl solution are shown in Figure 11. It can be seen that no serious collapse of platelet structure is observed after 7 days immersion and most of the structure remained almost intact and was shown to be similar to the as-prepared original structure (Figure 11a,b). However, due to the longtime contact, dissolution of some platelets of MgAl–LDH structure was observed at some points Strong Cl^- signals in EDS analysis (inset of Figure 11b) were observed after contact with chloride solution, so we can say that the MgAl–LDH film exhibited anion exchange capability by absorbing Cl^- while the XRD analysis also showed that the interlayer of the MgAl–LDH was able to retain Cl^- inside the structure. The LDH film anion-exchange behavior on the aluminum alloy in the chloride solution can be generally expressed as follows [27]:

$$LDH\text{-}NO_3 \quad + 2Cl^- \rightarrow LDH\text{-}2Cl^- + NO_3{}^- \tag{2}$$

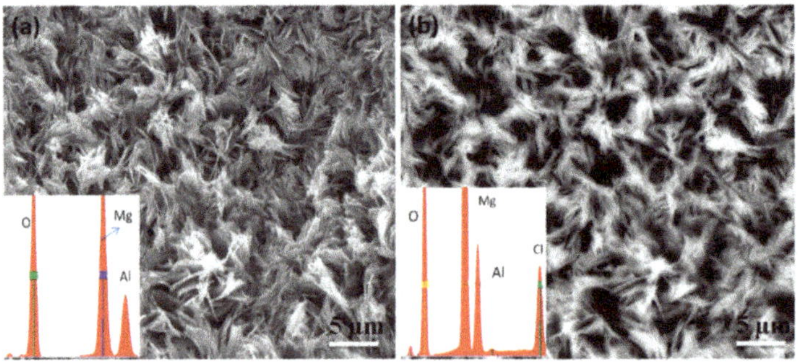

Figure 11. SEM images of (**a**) original MgAl LDH coating, 60 °C–24 h sample and (**b**) an immersed sample soaked in 0.1 M NaCl solution for 7 days (Insets are the corresponding EDS spectra).

5. Conclusions

The finding demonstrated that the reaction temperature and crystallization times have an influential effect on the growth mechanism of layered double hydroxide which ultimately has a strong impact on its corrosion resistance behavior. Well-grown uniformed structures were obtained at 80 and 100 °C with crystallization times of 18 and 24 h while the thickness of the obtained films lay in the range of 24–74 µm, which also imparted a significant influence on the corrosion resistance properties. The results from polarization curves confirmed that MgAl LDH stands as an improved corrosion protection thin film, which demonstrates significantly high impedance and low corrosion current density as compared to the substrate. The thicker coatings (>30 μm, obtained on treating the substrate at 100 °C for 18 and 24 h) showed a remarkable decrease of the corrosion current (3–4 orders of magnitude) and a remarkable increase in low-frequency impedance ($|Z|_{0.01}$) compared to the bare substrate. However, the long-term durability of the investigated coating is still an issue as after about one week of immersion in the 0.1 M NaCl solution, the impedance in the low-frequency range showed a decrease of a few orders of magnitude. Compared to the bare substrate the investigated coatings were revealed to be in any case quite protective compared to the bare AA6082 substrate.

Author Contributions: Conceptualization, M.F.; Data curation, M.A.I.; Formal analysis, M.A.I.; Funding acquisition, M.F.; Investigation, M.A.I.; Methodology, M.F. and M.A.I.; Supervision, M.F.; Project administration, M.F.; visualization, M.A.I., Writing—original draft, M.A.I.; Writing—review & editing, M.F.

Funding: This research received no external funding.

Conflicts of Interest: The authors declare no conflict of interest.

References

1. Davis, J.R. *Corrosion of Aluminium and Aluminium Alloys*; ASM International: Novelty, OH, USA, 1999.
2. Revie, R.W. *Corrosion and Corrosion Control: An Introduction to Corrosion Science and Engineering*, 4th ed.; John Wiley & Sons: New York, NY, USA, 2008.
3. Xhanari, K.; Finšgar, M. Organic corrosion inhibitors for aluminum and its alloys in chloride and alkaline solutions: A review. *Arab. J. Chem.* **2016**, in press. [CrossRef]
4. Kumar, N.; Rao, P.N.; Jayaganthan, R.; Brokmeier, H.G. Effect of cryorolling and annealing on recovery, recrystallisation, grain growth and their influence on mechanical and corrosion behaviour of 6082 Al alloy. *Mater. Chem. Phys.* **2015**, *165*, 177–187. [CrossRef]
5. Liang, W.J.; Rometsch, P.A.; Cao, L.F.; Birbilis, N. General aspects related to the corrosion of 6xxx series aluminium alloys: Exploring the influence of Mg/Si ratio and Cu. *Corros. Sci.* **2013**, *76*, 119–128. [CrossRef]
6. Larsen, M.H.; Walmsley, J.C.; Lunder, O.; Mathiesen, R.H.; Nisancioglu, K. Intergranular corrosion of copper-containing AA6xxx AlMgSi aluminum alloys. *J. Electrochem. Soc.* **2008**, *155*, C550–C556. [CrossRef]
7. Zubillaga, O.; Cano, F.J.; Azkarate, I.; Molchan, I.S.; Thompson, G.E.; Skeldon, P. Anodic films containing polyaniline and nanoparticles for corrosion protection of AA2024T3 aluminium alloy. *Surf. Coat. Technol.* **2009**, *203*, 1494–1501. [CrossRef]
8. Boisier, G.; Lamure, A.; Pébère, N.; Portail, N.; Villatte, M. Corrosion protection of AA2024 sealed anodic layers using the hydrophobic properties of carboxylic acids. *Surf. Coat. Technol.* **2009**, *203*, 3420–3426. [CrossRef]
9. Johansen, H.D.; Brett, C.M.; Motheo, A.J. Corrosion protection of aluminium alloy by cerium conversion and conducting polymer duplex coatings. *Corros. Sci.* **2012**, *63*, 342–350. [CrossRef]
10. Shi, H.; Han, E.H.; Liu, F.; Kallip, S. Protection of 2024-T3 aluminium alloy by corrosion resistant phytic acid conversion coating. *Appl. Surf. Sci.* **2013**, *280*, 325–331. [CrossRef]
11. Schäfer, H.; Stock, H.R. Improving the corrosion protection of aluminium alloys using reactive magnetron sputtering. *Corros. Sci.* **2005**, *47*, 953–964. [CrossRef]
12. Diesselberg, M.; Stock, H.R.; Mayr, P. Corrosion protection of magnetron sputtered TiN coatings deposited on high strength aluminium alloys. *Surf. Coat. Technol.* **2004**, *177–178*, 399–403. [CrossRef]
13. Arenas, M.A.; Conde, A.; de Damborenea, J.J. Effect of acid traces on hydrothermal sealing of anodising layers on 2024 aluminium alloy. *Electrochim. Acta* **2010**, *55*, 8704–8708. [CrossRef]
14. Venugopal, A.; Panda, R.; Manwatkar, S.; Sreekumar, K.; Krishna, L.R.; Sundararajan, G. Effect of micro arc oxidation treatment on localized corrosion behaviour of AA7075 aluminum alloy in 3.5% NaCl solution. *Trans. Nonferrous Met. Soc. China* **2012**, *22*, 700–710. [CrossRef]
15. Dalmoro, V.; dos Santos, J.H.; Armelin, E.; Alemán, C.; Azambuja, D.S. A synergistic combination of tetraethylorthosilicate and multiphosphonic acid offers excellent corrosion protection to AA1100 aluminum alloy. *Appl. Surf. Sci.* **2013**, *273*, 758–768. [CrossRef]
16. Voevodin, N.N.; Kurdziel, J.W.; Mantz, R. Corrosion protection for aerospace aluminum alloys by modified selfassembled nanophase particle (MSNAP) sol–gel. *Surf. Coat. Technol.* **2006**, *201*, 1080–1084. [CrossRef]

17. Qu, J.E.; Geng, C.H.E.N.; Wang, H.R.; Nie, D.J. Effect of water content on corrosion inhibition behavior of self-assembled TDPA on aluminum alloy surface. *Trans. Nonferrous Met. Soc. China* **2013**, *23*, 3137–3144. [CrossRef]
18. Zou, K.; Wang, X.; Liu, W.; Zhao, Y. Preparation and characterization of Ce−silane−ZrO_2 composite coatings on 1060 aluminum. *Trans. Nonferrous Met. Soc. China* **2014**, *24*, 1474–1480.
19. Lutz, A.; van den Berg, O.; Van Damme, J.; Verheyen, K.; Bauters, E.; De Graeve, I.; Du Prez, F.; Terryn, H. A shape-recovery polymer coating for the corrosion protection of metallic surfaces. *ACS Appl. Mater. Interfaces* **2014**, *7*, 175–183. [CrossRef]
20. Ates, M. A review on conducting polymer coatings for corrosion protection. *J. Adhes. Sci. Technol.* **2016**, *30*, 1510–1536. [CrossRef]
21. Leroux, F.; Taviot-Guého, C. Fine tuning between organic and inorganic host structure: New trends in layered double hydroxide hybrid assemblies. *J. Mater. Chem.* **2005**, *15*, 3628–3642. [CrossRef]
22. Evans, D.G.; Duan, X. Preparation of layered double hydroxides and their applications as additives in polymers, as precursors to magnetic materials and in biology and medicine. *Chem. Commun.* **2006**, *5*, 485–496. [CrossRef]
23. Williams, G.R.; O'Hare, D. Towards understanding, control and application of layered double hydroxide chemistry. *J. Mater. Chem.* **2006**, *16*, 3065–3074. [CrossRef]
24. Lei, X.D.; Yang, L.; Zhang, F.Z.; Evans, D.G.; Duan, X. Synthesis of oriented layered double hydroxide thin films on sulfonated polystyrene substrates. *Chem. Lett.* **2005**, *34*, 1610–1611. [CrossRef]
25. Chen, H.; Zhang, F.; Fu, S.; Duan, X. In situ microstructure control of oriented layered double hydroxide monolayer films with curved hexagonal crystals as superhydrophobic materials. *Adv. Mater.* **2006**, *18*, 3089–3093. [CrossRef]
26. Wang, L.; Li, C.; Liu, M.; Evans, D.G.; Duan, X. Large continuous, transparent and oriented self-supporting films of layered double hydroxides with tunable chemical composition. *Chem. Commun.* **2007**, *2*, 123–125. [CrossRef] [PubMed]
27. Zhang, F.; Zhang, C.; Song, L.; Zeng, R.; Liu, Z.; Cui, H. Corrosion of in-situ grown MgAl–LDH coating on aluminum alloy. *Trans. Nonferrous Met. Soc. China* **2015**, *25*, 3498–3504. [CrossRef]
28. Lin, J.-K.; Jeng, K.-L.; Uan, J.-Y. Crystallization of a chemical conversion layer that forms on AZ91D magnesium alloy in carbonic acid. *Corros. Sci.* **2011**, *53*, 3832–3839. [CrossRef]
29. Wang, X.; Li, L.; Xie, Z.-H.; Yu, G. Duplex coating combining layered double hydroxide and 8-quinolinol layers on Mg alloy for corrosion protection. *Electrochim. Acta* **2018**, *283*, 1845–1857. [CrossRef]
30. Wang, L.; Zhang, K.; He, H.; Sun, W.; Zong, Q.; Liu, G. Enhanced corrosion resistance of MgAl hydrotalcite conversion coating on aluminum by chemical conversion treatment. *Surf. Coat. Technol.* **2013**, *235*, 484–488. [CrossRef]
31. Liu, Z.; Ma, R.; Osada, M.; Iyi, N.; Ebina, Y.; Takada, K.; Sasaki, T. Synthesis, anion exchange, and delamination of Co–Al layered double hydroxide: Assembly of the exfoliated nanosheet/polyanion composite films and magneto-optical studies. *J. Am. Chem. Soc.* **2006**, *128*, 4872–4880. [CrossRef]
32. Okamoto, K.; Iyi, N.; Sasaki, T. Factors affecting the crystal size of the MgAl–LDH (layered double hydroxide) prepared by using ammonia-releasing reagents. *Appl. Clay Sci.* **2007**, *37*, 23–31. [CrossRef]
33. Wu, L.; Zheng, Z.; Pan, F.; Tang, A.; Zhang, G.; Liu, L. Influence of Reaction Temperature on the Controlled Growth of Mg–Al LDH Film. *Int. J. Electrochem. Sci.* **2017**, *12*, 6352–6364. [CrossRef]
34. Cullity, B.D. *Elements of X-ray Diffraction*; Pearson Education: London, UK, 2014.
35. Iqbal, M.A.; Fedel, M. The effect of the surface morphologies on the corrosion resistance of in situ growth MgAl–LDH based conversion film on AA6082. *Surf. Coat. Technol.* **2018**, *352*, 166–174. [CrossRef]
36. Aisawa, S.; Hirahara, H.; Uchiyama, H.; Takahashi, S.; Narita, E. Synthesis and thermal decomposition of Mn–Al layered double hydroxides. *J. Solid State Chem.* **2002**, *167*, 152–159. [CrossRef]
37. Wu, Q.; Olafsen, A.; Vistad, Ø.B.; Roots, J.; Norby, P. Delamination and restacking of a layered double hydroxide with nitrate as counter anion. *J. Mater. Chem.* **2005**, *15*, 4695–4700. [CrossRef]
38. Kloprogge, J.T.; Frost, R.L. Fourier transform infrared and Raman spectroscopic study of the local structure of Mg-, Ni-, and Co-hydrotalcites. *J. Solid State Chem.* **1999**, *146*, 506–515. [CrossRef]
39. Ni, Z.M.; Xia, S.J.; Fang, C.P.; Wang, L.G.; Hu, J. Synthesis, characterization and thermal property of Cu/Co/Mg/Al hydrotalcite like compounds. *Rare Met. Mater. Eng.* **2008**, *37*, 634–637.

40. Beving, D.E.; McDonnell, A.M.; Yang, W.; Yan, Y. Corrosion resistant high-silica-zeolite MFI coating one general solution formulation for aluminum alloy AA-2024-T3, AA-5052-H32, AA-6061-T4, and AA-7075-T6. *J. Electrochem. Soc.* **2006**, *153*, B325–B329. [CrossRef]
41. Tedim, J.; Kuznetsova, A.; Salak, A.N.; Montemor, F.; Snihirova, D.; Pilz, M.; Zheludkevich, M.L.; Ferreira, M.G.S. Zn–Al layered double hydroxides as chloride nanotraps in active protective coatings. *Corros. Sci.* **2012**, *55*, 1–4. [CrossRef]
42. Zhang, F.; Liu, Z.G.; Zeng, R.C.; Li, S.Q.; Cui, H.Z.; Song, L.; Han, E.H. Corrosion resistance of Mg–Al–LDH coating on magnesium alloy AZ31. *Surf. Coat. Technol.* **2014**, *258*, 1152–1158. [CrossRef]
43. Zhang, C.; Luo, X.; Pan, X.; Liao, L.; Wu, X.; Liu, Y. Self-healing Li–Al layered double hydroxide conversion coating modified with aspartic acid for 6N01 Al alloy. *Appl. Surf. Sci.* **2017**, *394*, 275–281. [CrossRef]
44. Wang, F.; Guo, Z. In situ growth of durable superhydrophobic Mg–Al layered double hydroxides nanoplatelets on aluminum alloys for corrosion resistance. *J. Alloy. Compd.* **2018**, *767*, 382–391. [CrossRef]
45. Lin, K.; Luo, X.; Pan, X.; Zhang, C.; Liu, Y. Enhanced corrosion resistance of LiAl–layered double hydroxide (LDH) coating modified with a Schiff base salt on aluminum alloy by one step in situ synthesis at low temperature. *Appl. Surf. Sci.* **2019**, *463*, 1085–1096. [CrossRef]
46. Tedim, J.; Zheludkevich, M.L.; Bastos, A.C.; Salak, A.N.; Lisenkov, A.D.; Ferreira, M.G.S. Influence of preparation conditions of layered double hydroxide conversion films on corrosion protection. *Electrochim. Acta* **2014**, *117*, 164–171. [CrossRef]

coatings

MDPI

Article

Inclusion of 5-Mercapto-1-Phenyl-Tetrazole into β-Cyclodextrin for Entrapment in Silane Coatings: An Improvement in Bronze Corrosion Protection

Cecilia Monticelli [1,*], Giancarlo Fantin [2], Graziano Di Carmine [2], Federica Zanotto [1] and Andrea Balbo [1]

[1] Centro di Studi sulla Corrosione e Metallurgia "A. Daccò", Dipartimento di Ingegneria, Università di Ferrara, 44121 Ferrara, Italy
[2] Dipartimento di Scienze Chimiche e Farmaceutiche, Università di Ferrara, 44121 Ferrara, Italy
* Correspondence: mtc@unife.it; Tel.: +39-0532-455136

Received: 23 July 2019; Accepted: 8 August 2019; Published: 10 August 2019

Abstract: The corrosion protection of coatings can be reinforced by the addition of entrapped corrosion inhibitors. β-cyclodextrin (β-CD) can form inclusion complexes with small inhibiting organic molecules that, when entrapped in coatings, allow the inhibitor release and adsorption at corrosion initiation sites. In this paper, several Nuclear Magnetic Resonance (NMR)-based experiments (e.g., Complexation-Induced Shifts (CIS), NMR titration, Diffusion-Ordered Spectroscopy (DOSY)) were performed to study the stability and geometry of a complex formed by β-cyclodextrin with 5-mercapto-1-phenyl-tetrazole (MPT). The complex was also detected by Electrospray Ionization (ESI) mass spectrometry and characterized by Fourier Transform Infrared (FTIR) spectra. Its influence on the protectiveness of a silane coating against bronze corrosion was evaluated in plain (AR) and concentrated (ARX10) synthetic acid rain, under different exposure conditions. In particular, the time evolution of the polarization resistance values during 20 days in ARX10 and the polarization curves recorded at the end of the immersions evidenced a higher protectiveness of the coating with the β-CD–MPT complex in comparison to that containing only MPT or only β-CD. The cyclic AR spray test carried out on coated bronze coupons with cross-cut scratches evidenced the absence of underfilm corrosion starting from the scratches only in the complex-containing coating.

Keywords: corrosion protection; smart coatings; inclusion complex; complex stability; bronze

1. Introduction

The corrosion protection of an organic coating depends largely on the intrinsic barrier properties of the polymeric film towards oxygen, water, and aggressive species, but can be reinforced by entrapped corrosion inhibitors [1–3]. In the case of bronze artworks exposed outdoors, commercial coatings such as Incralac® and Soter® contain benzotriazole (BTA) that operates in the dual functions of an inhibitor of bronze corrosion and an anti-UV additive [4]. In these coatings, the inhibitor dissolves in the electrolyte (in the rain) and penetrates through the coating, so producing an inhibited solution that exerts a protective action at the coating/metal interface. In general, the positive influence of direct inhibitor addition in coatings may be limited by solubility problems, by a decrease in barrier properties and adherence to the substrate, and by the rapid leaching of small inhibitor molecules induced by rainfalls, so determining a fast drop of the overall protection performance [5].

In recent years, great efforts were devoted to overcoming these problems and increasing the coating durability by encapsulating corrosion inhibitors in coatings through the adoption of suitable carriers, which make them more compatible with the coating network [6,7]. Different carrier types were investigated, such as inhibitor-filled porous particles, or nanocapsules, coated by polyelectrolytes [8–10] or layered anion-exchange particles containing intercalated corrosion inhibiting anionic species [11]. The former polyelectrolyte-coated carriers can release inhibitor molecules due to local variations of solution pH linked to the onset of corrosion phenomena, while the latter carriers with intercalated inhibiting anionic species limit the access of aggressive anions, such as chlorides, to the metal substrate and release corrosion inhibiting species by ion-exchange phenomena.

Beside these carriers, cyclodextrins (CD) were also used to host corrosion inhibitors due to their complexing capability towards small organic molecules with corrosion inhibition properties. Cyclodextrins (CDs) [12] are water-soluble macrocyclic oligosaccharides consisting of at least six α -D-glucopyranose units linked via α (1–4) glycosidic bonds. The most common ones contain 6, 7 and 8 glucopyranose residues and are known as α-, β-, and γ-CD, respectively. CDs are among the most widely used host molecules thanks to their unique and specific structure that creates an internal cavity that is less hydrophilic than the external aqueous environment and therefore able to accommodate a large variety of hydrophobic molecules inside it [13]. The strong propensity of CDs to form inclusion complexes has been exploited not only for the production of smart coatings for corrosion protection, but also in many other fields of science, for example, as drug transport systems, [14], to increase the solubility of some chemical species [15], in separation technology [16], and in other areas [17].

Addition of complexes of α- and β-CDs with dibenzylthiourea (DBT) in acid solutions improved the corrosion resistance of carbon steel due to the enhanced solubility of the complexes in comparison to that of DBT alone [18]. The controlled release of BTA from β-CD–BTA complexes was investigated in order to achieve effective bronze corrosion protection in chloride solutions. The use of the complex instead of pure BTA was intended to reduce the toxicity of the additives used for corrosion protection [19]. Moreover, complexes of β-CD and γ-CD with mercaptobenzothiazole (MBT), mercaptobenzimidazole (MBI), mercaptobenzimidazole sulfonate, and thiosalicylic acid [20–23] were incorporated in coatings and improved the corrosion resistance of Al alloys and zinc. In fact, these complexes represent a reservoir of corrosion inhibitor molecules that, at the onset of corrosion phenomena and in correspondence of regions of coating delamination, tend to adsorb on the metal surfaces, shifting the complexation equilibrium towards the release of further inhibitor molecules.

In this research, the effective formation and the stability of a complex between β-CD and 5-mercapto-1-phenyl-tetrazole (MPT) (Figure 1) were assessed. Among non-toxic corrosion inhibitors [24–26], MPT was chosen because of its outstanding inhibiting properties towards copper and bronze corrosion [27–29]. Its complex with β-CD was incorporated in a 3-mercapto-propyl-trimethoxy-silane (PropS-SH) coating, and the protectiveness of the obtained coating was assessed on bronze during both immersions in concentrated acid rain and exposures to alternated acid rain spray. These tests were also performed on plain silane coating and coatings with only β-CD or only MPT additions. For the continuous immersion tests, thin sub-micrometric silane films prepared by the dip coating method were used, in order to better differentiate their protectiveness. During the alternated acid rain spray test, the self-healing properties were assessed on thicker coatings (about 5 μm) produced by spraying after introduction of cross cut scratches. Spraying and brushing are the application methods most commonly adopted by restorers [30] and produce coatings more representative of those actually applied in the field.

It is important to stress that according to occupational hazard tests, the silane formulations here investigated were less toxic to restorers than Incralac® (as both pure product and ready to use 30% solution in nitro diluent) [31].

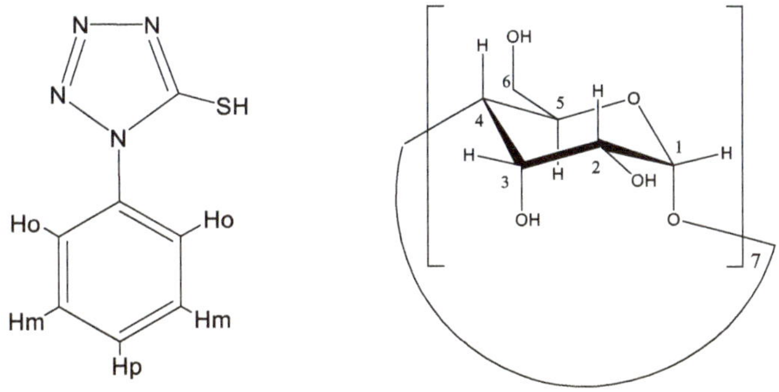

(**a**) (**b**)

Figure 1. Molecular structure of (**a**) 5-mercapto-1-phenyl-tetrazole (MPT) and (**b**) β-cyclodextrin (β–CD). β–CD is a CD type constituted by 7 glucopyranoside units linked by 1,4 glycosydic bonds.

2. Materials and Methods

2.1. *Chemicals, Aggressive Environment and Alloy*

All reagents and solvents used in this study were purchased from commercial sources. In particular, the chemicals used for the inhibitor complex and the coating production were β-cyclodextrin (β-CD, ≥ 97% purity), 5-mercapto-1-phenyl-tetrazole (MPT, 98% purity), and 3-mercapto-propyl-trimethoxy-silane (PropS-SH, purity 95%), all purchased from Sigma-Aldrich (Darmstadt, Germany).

The coating protectiveness was tested on as-cast bronze with composition 91.9 Cu, 2.4 Sn, 1.0 Pb, Zn 2.9, 0.8 Sb, wt.%, and a microstructure reproducing those of Renaissance bronze artefacts with cored dendrites of Cu-solid solution characterized by Sn and Sb local enrichment and also including Pb globules in the interdendritic spaces, as reported in previous papers [29,31–33].

Concerning the environments where the coating protectiveness was assessed, the cyclic acid rain (AR) spray test was performed using a synthetic AR, prepared with Sigma-Aldrich ACS reagents, according to the recipe reported in [34] and containing the following ion concentrations: SO_4^{2-} 1.90 mg· L^{-1}, Cl^- 1.27 mg· L^{-1}, NO_3^- 4.62 mg· L^{-1}, CH_3COO^- 0.23 mg· L^{-1}, $HCOO^-$ 0.05 mg· L^{-1}, NH_4^+ 1.05 mg· L^{-1}, Ca^{2+} 0.34 mg· L^{-1}, Na^+ 0.53 mg· L^{-1}, and pH 4.35. During the electrochemical tests, accelerated corrosion conditions were obtained by using tenfold concentrated AR (ARX10, pH = 3.3) at 30 °C.

2.2. *[β-CD–MPT] Complex Stability Analysis*

2.2.1. Nuclear Magnetic Resonance (NMR) Measurements

The NMR spectra were recorded in D_2O solution using 5 mm tubes, at 296 K, with a Varian Mercury Plus 400 (Palo Alto, CA, USA), operating at 400 (^1H) and 100 MHz (^{13}C), respectively. The chemical shifts were referenced to the DOH signal: δ (H) 4.65 ppm. The 1D-Rotating frame Overhauser effect spectroscopy (ROESY) NMR spectra were acquired using standard pulse sequences from the Varian library. The relaxation delay between successive pulse cycles was 1.0 s.

The Diffusion-Ordered Spectroscopy (DOSY) experiments were performed using the Dosy Bipolar Pulsed Pair STimulated Echo (DBPPSTE) pulse sequence [35] from the Varian library, using 15 different gradient values varying from 2% to 95% of the maximum gradient strength. A 500 ms diffusion time was chosen, and the gradient length was set to 2.0 ms. The analysis of all NMR spectra was performed with MestreNova (by Mestrelab Research, S.L., Santiago de Compostela, Spain), version: 6.0.2–5475 and for the DOSY analysis, the Baysian DOSY transform from MestreNova, version: 6.0.2–5475 was used.

2.2.2. H NMR Titration

The following two solutions were prepared in D_2O: Solution A: 2.8 mM MPT. Solution B: 2.8 mM MPT and 12.0 mM β-CD. A 0.8 mL aliquot of solution A was placed in a 5 mm NMR tube. A measured amount of solution B was added, changing the molar fraction of the host to about 0, 0.39, 0.71, 0.98, 1.21, 1.68, 2.02 and 2.29. Spectra were recorded after each addition. The chemical shift variation of the guest signals was collected, and the binding constants β (as log K) were calculated by the curve fitting method [36] using the commercial HypNMR2008 [37] program (details are given in the Supplementary Information (SI) file).

2.2.3. Electrospray Ionization (ESI) Mass Spectra

ESI mass spectra were obtained using an LCQ Duo (ThermoQuest, San Jose, CA, USA) in negative-ion mode. Instrumental parameters: capillary voltage –10 V, spray voltage 4.50 kV, mass scan range was from m/z 100 to 2000 amu, for 30,000 ms scan time; N^2 was used as the sheath gas. The samples were injected into the spectrometer through a syringe pump at a constant flow rate of 8 mL/min.

2.2.4. Fourier Transform Infra-Red (FTIR) Analysis

Diffuse reflectance FTIR spectra were recorded on β-CD–MPT complex powder and on MPT and β-CD powders, as references. The instrument used was a Thermo-Scientific Nicolet iS50 spectrometer (Thermo Fisher Scientific, Waltham, MA, USA), operating in dry CO_2-free air flow generated by a Balston 75-52 unit. It was equipped by a deuterated triglycine sulfate (DTGS) detector, which allowed for the investigation of the 4000–400 cm^{-1} wavenumber region with a resolution of 4 cm^{-1}.

2.3. *Silane Coating Production*

Silane hydrolysis was carried out by dissolving PropS-SH in a hydroalcoholic solution (90/5/5 v/v ethanol/water/PropS-SH), acidified to pH 4 by the addition of some drops of diluted sulphuric acid solution, according to the methodology refined in previous research works [29,38–40].

Plain PropS-SH coatings were directly produced from this solution after 24 h room-temperature ageing. The coating additives were introduced in the aged silane hydroalcoholic solution; in particular, 3 mL aqueous solutions of either 5.94 mM MPT or 5.94 mM β-CD or 5.94 mM β-CD + 5.94 mM MPT were added to 30 mL of silane solution, so that the final molar concentration of the additives in the coating formulations was 0.54 mM. These solutions were sonicated for 3 min and then applied to the substrate either by dip coating (1 h immersion and then fast withdrawal, reaching a final coating thickness of about 300 nm; for accelerated electrochemical tests) or by spraying (to reach a final constant coating specific weight of 6 ± 1 $g \cdot m^{-2}$ and thickness of 5 ± 1 μm; for cyclic AR spray tests). Finally, the coatings were cured for 24 h at 50 °C. This low temperature curing was compatible with the requirements for cultural heritage bronze artworks.

2.4. *Silane Coating Protectiveness*

Electrochemical tests were performed under accelerated corrosion conditions, that is in ARX10 (pH 3.3) at 30 °C, on thin dip-coated bronze electrodes. As a reference, tests were also carried out on bare bronze electrodes.

The evolution of corrosion conditions was monitored over 20 days of immersion by Electrochemical Impedance Spectroscopy (EIS, performed by a PARTSTAT 2273, from Ametek, Berwyn, PA, USA) tests performed at intervals, under the following experimental conditions: corrosion potential (E_{cor}) ± 10 mV rms, 10 kHz–1 mHz frequency range and 10 frequencies/decade. Polarization resistance (R_p) values were estimated from the spectra in the Nyquist form, as the difference between the limit of the real part of the impedance at frequency tending to 0 (R_p') and the solution resistance (R_s) value ($R_p = R_p' - R_s$) [40]. R_p values are inversely proportional to the corrosion currents (i_{cor}), as indicated by the Stern and Geary relationship [41]: $R_p = \frac{B}{i_{cor}}$, with B a constant depending on the Tafel slopes of the anodic and cathodic polarization curves. The time evolutions of average R_p and E_{cor} values were obtained from triplicate experiments.

Ohmic drop-compensated polarization curves were collected at the end of the 20 day immersion period. Separate anodic and cathodic potential scans, always starting from E_{cor}, were carried out at a rate of 0.1667 mV·s^{-1}. These tests were performed in triplicate and representative curves were reported.

The self-healing capability of PropS-SH coatings was assessed by exposing coated coupons with cross cut scratches to a cyclic AR spray test at 35 °C for 4 weeks. Each cycle consisted in 8 h spraying and 16 h waiting. During the test, the coupons were supported with an angle of 30° from the vertical. Micrographs documented the extent of the final corrosion attack.

3. Results

3.1. NMR Studies

To verify the effective formation of a host–guest complex between β-CD and MPT we mainly employed the NMR spectroscopy, already widely used to study inclusion complexes of CD [42–44].

The ^{1}H NMR spectrum in D_2O of a sample containing an equimolar amount of β-CD and MPT showed induced chemical shifts (Δδ) for all the protons of the host as well as for most of the guest protons (Figure 2). The complexation-induced shifts (CIS) observed for the internal protons of β-CD (H3, H5) were remarkably larger than those displayed by the external ones (H-1, H-2, and H-4) (Table 1), thus indicating that the guest likely interacts with the inner cavity of β-CD.

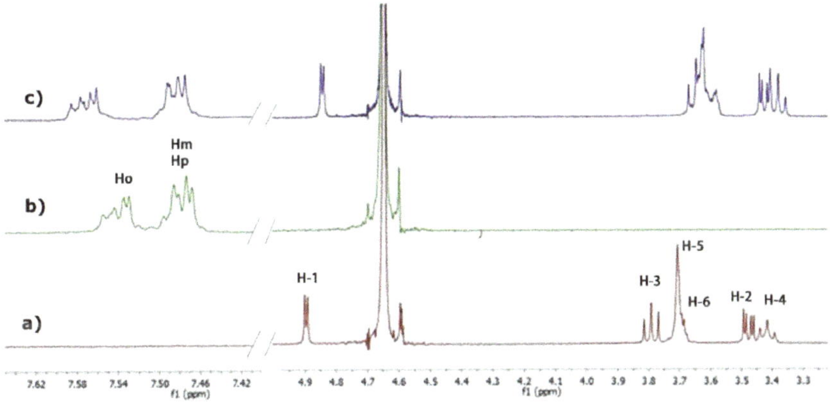

Figure 2. The ^{1}H NMR spectra in D_2O for solutions of (**a**) β-CD; (**b**) MPT; (**c**) equimolar amount of β-CD and MPT.

Table 1. The ^{1}H-NMR chemical shifts (δ, ppm) for H protons of β-CD alone, MPT alone, and complexation induced shifts (CIS = $\delta_{complex} - \delta_{free}$) of equimolar amounts of them in D_2O at 23 °C.

β-CD	**Protons**	**H-1**	**H-2**	**H-3**	**H-4**	**H-5**	**H-6**
δ alone		4.90	3.48	3.80	3.42	3.71	3.69
CIS		−0.06	−0.06	−0.15	−0.04	−0.09	−1.1
MPT	**Protons**	**H-o** [a]	**H-m** [a,b]	**H-p** [a,b]			
δ alone		7.54	7.48	7.48			
CIS		0.04	0.01	0.01			

[a] The center of the multiplet was taken into account; [b] overlapping signals.

The small CIS observed for the ortho-protons of the phenyl moiety of MPT (0.04 ppm, Table 1) was also particularly significant, indicating that a part of the phenyl ring of MPT entered into the cavity of β-CD. It is worthy of note that no new peaks appeared in the spectrum, signifying that the inclusion of MPT in β-CD is a fast exchange process that takes place on the NMR timescale.

In order to gather information on the sites of binding we carried out a series of monodimensional ROESY-1D experiments [45] that provided only a small nuclear Overhauser effect (NOE) on the inner H3 proton of β-CD when ortho-H phenyl protons of MPT were irradiated.

The experimental observations collected up to this point were compatible with the inclusion structure of Figure 3, which was in rapid equilibrium with the two separate molecules. In fact, the large CIS exhibited by inner protons H3 and H5 of β-CD upon the addition of MPT clearly indicated a deep insertion of the host into the hydrophobic cavity of the β-CD. However, the lack of dipolar interactions between the ortho- and meta-protons of the aromatic ring of MPT and the H5 proton of β-CD excluded the complete insertion of the phenyl moiety into the β-CD cavity. On the other hand, the weak rotating-frame Overhauser effect (ROE) existing between the ortho-protons of MPT and the H3 proton of β-CD suggested that a partial insertion of the phenyl moiety of MPT occurred at the wide rim of β-CD. Finally, the upfield shift of the H-6 protons of the β-CD (see Table 1) could be justified by the partial protrusion of the guest (tetrazole moiety) from the narrow rim of the β-CD (Figure 3) [46].

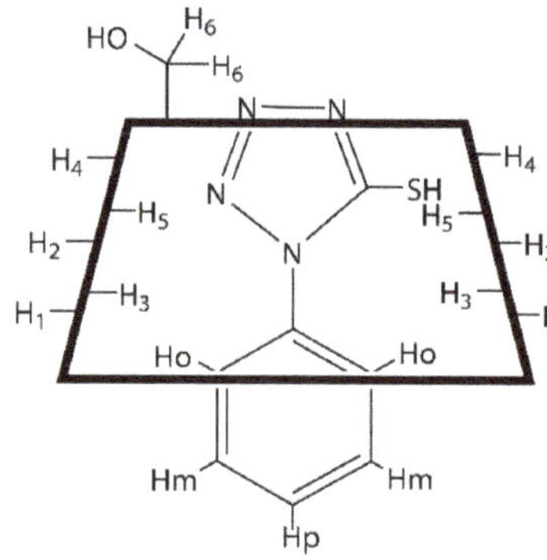

Figure 3. Proposed geometry for the inclusion of MPT into β-CD as deduced from the CIS and ROESY-1D experiments.

To investigate the strength of complexation we carried out a ^{1}H NMR titration of MPT with β-CD [36]. The procedure adopted in the titration allowed us to operate with a constant guest concentration ([MPT] = constant) during the whole experiment.

The binding isotherm relative to the ortho-protons of the phenyl moiety of MPT is depicted in Figure 4. The value of the stability constant β as log K_a of the β-CD–MPT inclusion complex was calculated by the curve fitting method [36], using the commercial HypNMR2008 [37] program (details are given in SI) and was found to be equal to 2.93 M^{-1} (K_a = 851 M^{-1}).

The 1:1 stoichiometry of the complexation adequately described the binding data obtained from the NMR titration and, on the other hand, the physically unrealistic binding parameters (some negative K_s) when the 1:2 or 2:1 models were applied to the NMR titration. This confirmed that the 1:1 stoichiometry was dominating in the investigated concentration range. The 1:1 stoichiometry of the complex was further confirmed by ESI-Mass Spectrometry (ESI-MS).

DOSY spectroscopy [37,42–47] experiments were also carried out to confirm, qualitatively and quantitatively, the intermolecular interactions between MPT and β-CD in solution. The DOSY technique allowed for the determination of the individual self-diffusion coefficients (D) in multicomponent systems that directly reflected the association behavior of the interacting species [48].

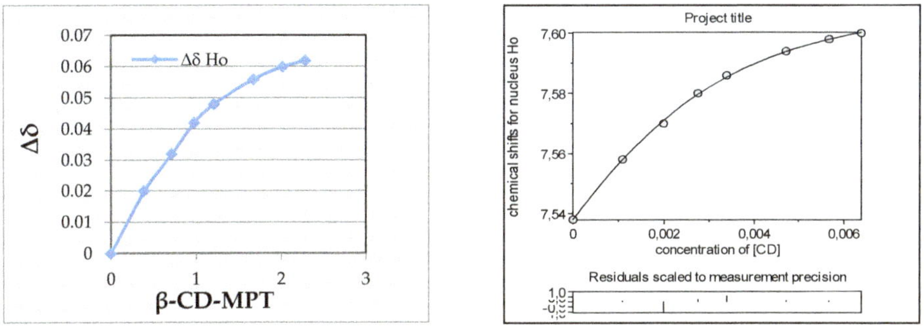

(a) (b)

Figure 4. The ^{1}H NMR titration of MPT with β-CD: (**a**) Chemical shift change of the ortho-protons of the phenyl moiety of MPT with increasing β-CD concentration. Positive values mean downfield shifts. The small complexation induced shift (CIS) observed for the ortho-protons of the phenyl moiety of MPT (0.04 ppm, see Table 1) was particularly significant, indicating that a part of the phenyl ring of MPT entered into the cavity of β-CD. (**b**) Curve-fitting analysis by the HypNMR2008 program.

The principle on which DOSY is based is very simple and can be summarized as follows: when the host and the guest are in the free state, they have their own diffusion coefficient that depends on their molecular weight and their shape. However, when they interact tightly together to form a complex, they behave as a single molecular entity and therefore should have the same diffusion coefficient [48,49].

Taking into account the fact that we are studying a rapid equilibrium on the NMR time scale between bound and free guest molecules, the observed (measured) diffusion coefficient (D_{obs}) is the weighted average of the free and bound diffusion coefficients (D_{free} and D_{bound}, respectively) and can therefore be used to calculate the bound fraction p by using the following Equation (1):

$$D_{obs} = p \cdot D_{bound} + (1-p) \cdot D_{free} \tag{1}$$

which can be rearranged to yield:

$$p = \frac{D_{free} - D_{obs}}{D_{free} - D_{bound}} \tag{2}$$

where p is the fraction of complexed substrate molecules.

After binding of a small guest molecule (MPT) to a large host molecule (β-CD) the diffusion coefficient of the host is not greatly perturbed, therefore, the diffusion coefficient of the host–guest complex can be assumed to be the same as that of the non-complexed host molecule [50].

Pseudo 2D DOSY spectra are shown in Figure 5. The f1 dimension represents the self-diffusion coefficient (D) and the f2 dimension reports the chemical shift. The f1 is specific for each molecule, so the protons belonging to the same molecule appear in the same f1 row. The diffusion coefficients (D) and the fraction of complexed MPT molecules (p) measured at 23 °C in D_2O are reported in Table 2.

As expected, the D value of encapsulated MPT ($4.87 \cdot 10^{-6}$ $cm^2 \cdot s^{-1}$) was lower than that of free MPT ($6.205 \cdot 10^{-6}$ $cm^2 \cdot s^{-1}$) (Table 2, Figure 5) thus proving that MPT is included in the β-CD cavity and diffuses more slowly. Recalling that the association constant, K_a, for a 1:1 host–guest equilibrium of the type H + G ⇄ HG is defined by:

$$K_a = \frac{[HG]}{[H][G]} \tag{3}$$

where [H], [G], and [HG] are the equilibrium concentrations of the free host, free guest, and complex, respectively, Equation (3) can be rewritten as a function of the molar fraction [48] as:

$$K_a = \frac{p}{(1-p)([H]_0 - p[G]_0)} \quad (4)$$

where $[H_0]$ and $[G_0]$ are the total concentrations of the host and guest, respectively.

Lastly, inserting the value of the molar fraction just obtained from the DOSY experiments in the above Equation (4) we can calculate K_a by using the single-point procedure [51,52]. The association constant measured in this way was 654 M^{-1} at 23 °C (Table 2), and although the value obtained by the single-point approximation method results in large uncertainty, it is consistent with the corresponding value of 851 M^{-1} estimated via NMR titration.

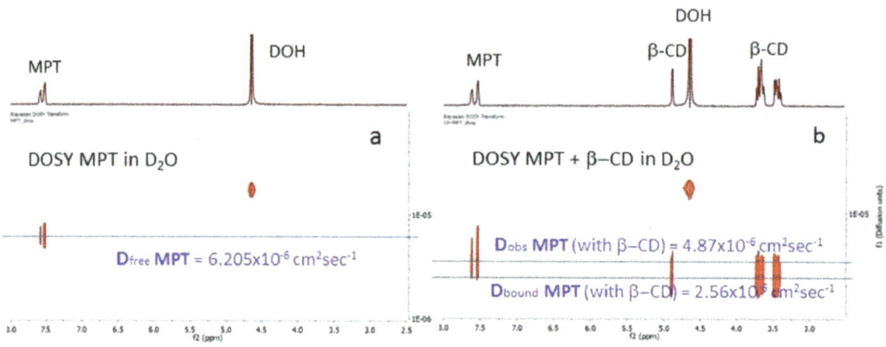

(a) (b)

Figure 5. Pseudo 2D DOSY spectra of MPT (**a**) 1.5 mM; (**b**) in the presence of β-CD 1.4 mM in D_2O, at 23 °C. (See Supplementary Information, SI, for details).

Table 2. Diffusion coefficients (D, 10^{-6} $cm^2 \cdot s^{-1}$) related to the itemized protons of MPT_{free} (MPT alone 1.5 mM), $MPT_{(+\ \beta\text{-CD})}$ (MPT 1.5 mM with β-CD 1.4 mM), and $\beta\text{-CD}_{(+\ MPT)}$ (MPT 1.5 mM with β-CD 1.4 mM), fraction of complexed MPT and β-CD molecules (p) and association constant for the β-CD–MPT complex (K_a).

Chemical Species	Proton (ppm)	$D^{a,b}$	$D_{average}$	p_{bound}	K_a (M^{-1})
$\mathbf{MPT_{(free)}}$	H_o (7.59)	6.22	6.205	0	–
	H_m/H_p (7.51) [c]	6.19	–	–	–
$\mathbf{MPT_{(+\ \beta\text{-CD})}}$	H_o (7.62)	4.88	4.87	0.36	654
	H_m/H_p (7.52) [c]	4.86	–	–	–
β-CD $_{(+\ MPT)}$	H_m/H_p (7.52) [c]	2.256	2.56	0.36	–
	$H_3/H_5/H_6$ (3.67) [c]	2.57	–	–	–
	H_2/H_4 (3.43) [c]	2.55	–	–	–

[a] in D_2O at 23 °C; [b] estimated errors <5%; [c] overlapped signals.

3.2. *ESI Mass Spectra*

In order to provide further confirmation of the formation of the β-CD–MPT inclusion complex, some ESI-MS experiments were conducted in aqueous solutions containing MPT and β-CD in the ratio 1:1, 1:2, and 2:1. In all cases, a base peak at m/z 1311 corresponding to a 1:1 host–guest complex was detected (Figure 6).

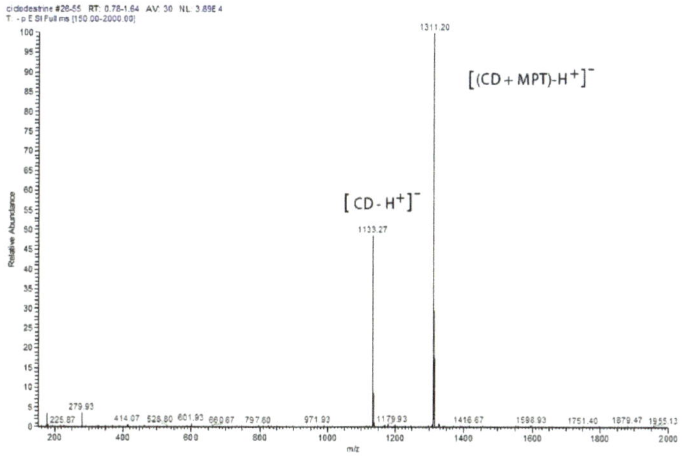

Figure 6. Electrospray negative-ion mass spectrum (ESI-MS) of a 1:1 β-CD–MPT aqueous solution that revealed a base peak corresponding to the 1:1 host–guest complex at m/z 1311 and a peak at m/z 1133 (50%) relative to uncomplexed β-CD.

3.3. FTIR Spectra

The diffuse reflectance FTIR spectrum of the solid β-CD–MPT complex precipitated from equimolar β-CD and MPT aqueous solutions was recorded and compared to those obtained on pure MPT and β-CD. Figure 7 clearly shows that the spectrum of the complex almost completely overlapped that of β-CD, but two extra peaks at 1492 cm^{-1} and 1593 cm^{-1} occurred (as evidenced in the enlarged inset), which corresponded to intense bands of the MPT molecule and did not appear in the β-CD spectrum. According to X.R. Ye et al. [53], both peaks were connected to the C–C stretching of the phenyl ring in MPT and the former also corresponded to N–H bending. These spectra further confirmed the complex formation.

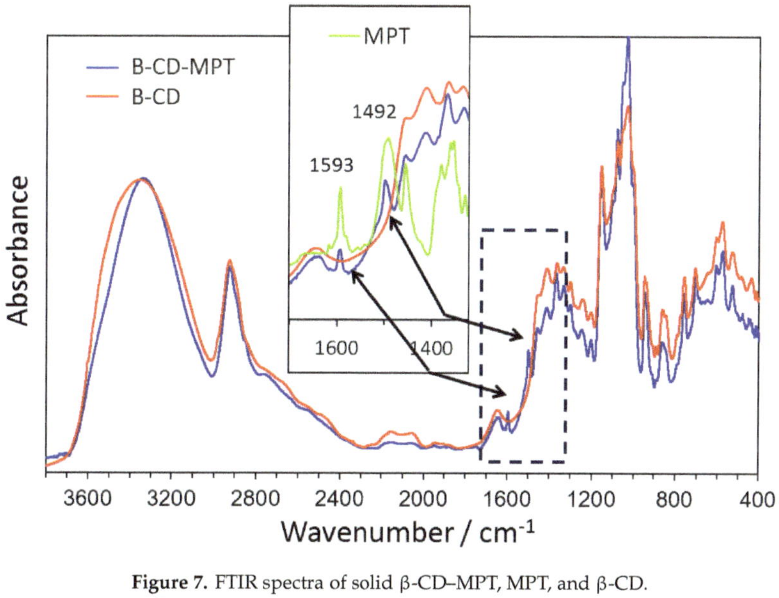

Figure 7. FTIR spectra of solid β-CD–MPT, MPT, and β-CD.

3.4. Electrochemical Tests

These tests were performed on dip coated bronze electrodes in ARX10, because the association of the obtained thin coatings and the higher aggressiveness of the concentrated solution allowed for the differentiation of the coating protectiveness within the 20 day immersion period.

Figure 8 collects the time evolution of R_p and E_{cor} values in this environment at 30 °C. For bare electrodes, high initial R_p values (9.1 kohm·cm^2) were obtained, which decreased quickly to about 1 kohm·cm^2 for immersions longer than 1 h and then increased again up to 4.6 kohm·cm^2 towards the end of the immersion period (Figure 8a). PropS-SH coatings showed much higher initial R_p values (almost 700 kohm·cm^2), which decreased by about 1 order of magnitude during the 20 days of immersion due to the slow penetration of the aggressive solution through the silane network. The addition of MPT to the silane solution determined rather low and constant R_p values, close to 20 kohm·cm^2. This behavior was not investigated but it is plausible that MPT interfered with the coating reticulation and/or a surface competitive adsorption between free MPT molecules on one side and the silanol and thiol groups of silane coatings on the other occurred, so impairing the coating adherence and performance. Instead, some improvements were achieved with the addition of β-CD, which due to its hydroxyl groups was likely capable of reacting with the silanol groups, so contributing to the silane network formation. Finally, a clear progression was observed after β-CD–MPT complex addition with high and rather constant R_p values (around 1 Mohm·cm^2). This suggests that beside the positive effect of β-CD, the release of MPT molecules from the β-CD cavity could also play an important role in corrosion inhibition.

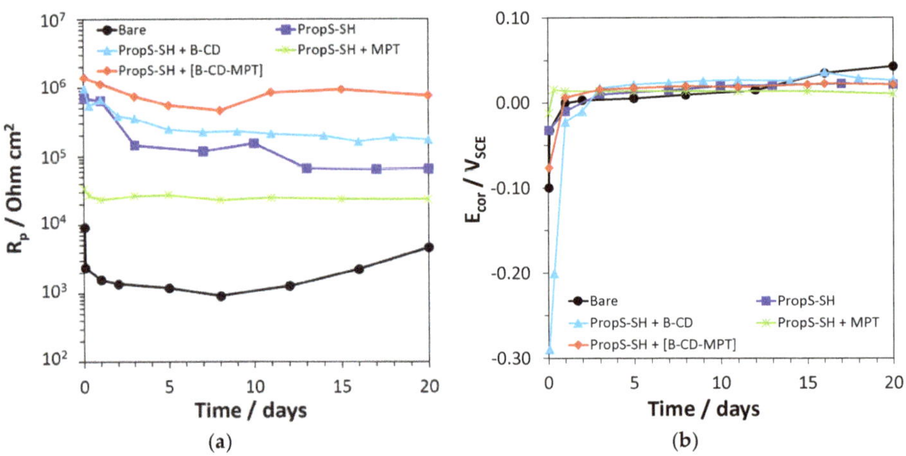

Figure 8. Polarization resistance (R_p) (**a**) and corrosion potentials (E_{cor}) values (**b**) measured during 20 days of immersion of bare and coated bronze electrodes in concentrated acid rain (ARX10) at 30 °C.

The E_{cor} values on bare electrodes evolved from −0.100 V_{SCE} after 1 h immersion to about +0.043 V_{SCE} after 20 days (Figure 8b). From previous research [54], it was found that this trend was the consequence of the degradation of the protective surface air-formed oxide film during the first half of the immersion period, which stimulated the cathodic reaction, with a consequent E_{cor} shift towards nobler values. Then, in the second half of the immersion, the progressive accumulation of surface corrosion products induced a slight inhibition of the anodic process and further consequent E_{cor} ennoblement. No significant E_{cor} differences were detected in the presence of the coatings. In all cases, E_{cor} increased to a certain extent during the initial 2 or 3 days of immersion, likely due to the evolution in the coating reticulation [39]. Then, they reached values in the range from 0.011 to 0.030 V_{SCE}, independently of the corresponding R_p values.

The polarization curves recorded at the end of the 20 days of immersion on bare and coated electrodes are shown in Figure 9, while Table 3 reports the electrochemical parameters derived from these curves.

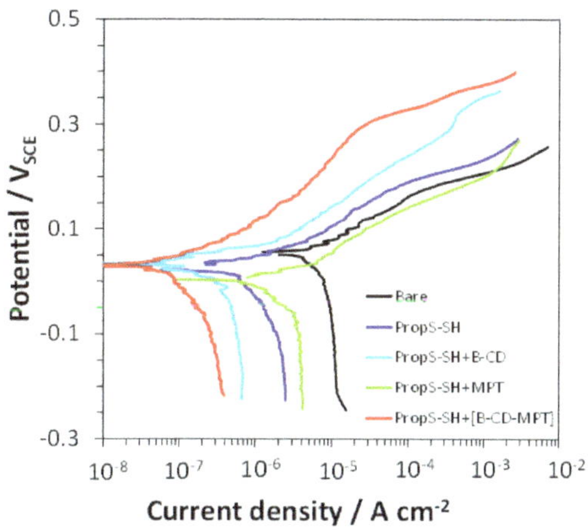

Figure 9. Polarization curves recorded on bare and silane (PropS-SH)-coated bronze specimens in the absence and in the presence of β-CD, MPT, or β-CD–MPT complex, after 20 days of immersion in ARX10 at 30 °C.

Table 3. Corrosion potentials (E_{cor}), corrosion current densities (i_{cor}), and anodic Tafel slopes (b_a) obtained on bare and coated bronze specimens after 20 days of immersion in ARX10 solution. Protection efficiencies (η) of silane coatings are also reported.

Coating Type	E_{cor}/V_{SCE}	$i_{cor}/\mu A \cdot cm^{-2}$	b_a/V	η/%
-	0.055	8.3	0.090	–
PropS-SH	0.024	1.3	0.092	84
PropS-SH + β-CD	0.030	0.70	0.088	92
PropS-SH + MPT	0.008	2.8	0.093	66
PropS-SH + β-CD-MPT	0.029	0.15	0.101	98

In particular, Table 3 collects the E_{cor} and corrosion current (i_{cor}) values, the anodic Tafel slopes, b_a, and the protection efficiency (η) of the coatings, evaluated by the formula:

$$\eta = \frac{i_{cor,b} - i_{cor,c}}{i_{cor,b}} \times 100 \quad (5)$$

where $i_{cor,b}$ and $i_{cor,c}$ are the corrosion currents evaluated on bare and coated electrodes, respectively. The cathodic Tafel slopes, b_c, which were not reported in Table 3, were generally much higher than b_a and close to infinity.

Figure 9 and Table 3 evidence that the coatings protect the underlying alloy from corrosion at different degrees. The plain PropS-SH coating mainly hindered the cathodic reaction, so determining a slight reactivation of E_{cor} in comparison to those of the bare electrodes, and afforded a final η of about 84%. In agreement with the R_p results, the addition of MPT to PropS-SH was detrimental to the coating protectiveness because it stimulated both the anodic and the cathodic reactions, suggesting a lower barrier effect of the coating and/or a lower surface adherence. In this case, the coating η value decreased down to 66% (Table 3). Conversely, the addition of β-CD or, even more, the β-CD–MPT complex in the coatings determined a decrease in the anodic and cathodic currents and induced η values of 92% and 98%, respectively. In contrast with MPT, the complex proved to be beneficial to the coating performances, likely due to a higher compatibility of the external β-CD surface with the silane network and to the specific inhibition afforded by released inhibitor molecules as evidenced by the following test.

3.5. Cyclic AR Spray Test

The self-healing capability of the silane coatings connected to the inhibitor release and adsorption at corrosion sites was evaluated during four weeks of exposure of the coated specimens with cross cut scratches to cyclic AR spray.

At the end of the test (Figure 10), the bronze coupons evidenced that the PropS-SH coating was rather protective at a distance from the scratches, but obviously did not avoid the substrate corrosion in scratched areas from which in fact the underfilm corrosion propagated. This corrosion form and coating delamination were less evident in the presence of β-CD, suggesting that this substance also increased the coating adherence, so improving the overall coating protectiveness. However, again, it did not prevent corrosion in the scratches. The addition of MPT in PropS-SH could not avoid the spread of corrosion attacks from the scratches and induced a significant surface color change (a brightening). Only β-CD–MPT complex addition significantly enhanced the substrate corrosion resistance and completely suppressed corrosion in the scratches and underfilm. This behavior suggested a self-healing capability of this coating type in the case of coating defects and mechanical damages, so prolonging the coating effectiveness.

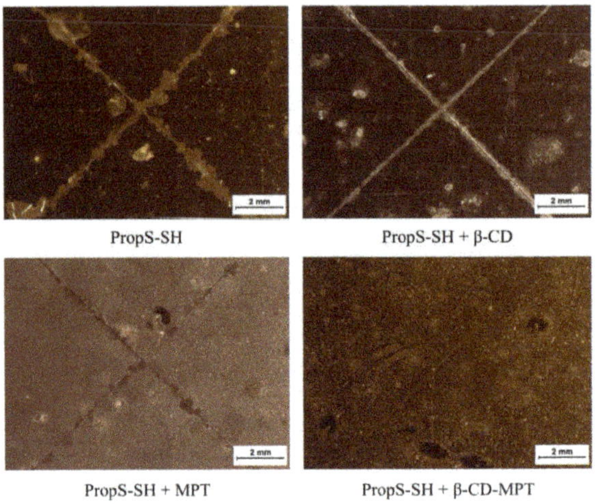

Figure 10. Surface aspect of cross-cut coupons at the end of a 4 week exposure to a cyclic acid rain (AR) spray test; only the PropS-SH coating containing the β-CD–MPT complex prevented the development of a corrosion attack starting from the scratches.

4. Conclusions

- The stability and the molecular structure of the host–guest β-CD–MPT complex was assessed by NMR, FTIR, and MS techniques.
- The analysis of the Complexation Induced Shifts suggested an inclusion structure of the complex, with partial insertion of the phenyl moiety of MPT at the wide rim of the hydrophobic β-CD cavity and partial protrusion of the tetrazole moiety of the inhibitor from the narrow rim of the host cavity.
- According to the DOSY experiments, the stability constant of the β-CD–MPT complex was 654 M^{-1}, in good agreement with the value of 851 M^{-1} obtained by curve fitting the NMR binding isotherm.

- The PropS-SH coating with entrapped β-CD–MPT complex exhibited an improved protectiveness (η = 98%) against bronze corrosion in comparison to plain PropS-SH (η = 84%) or PropS-SH containing only MPT (η = 66%) or β-CD (η = 92%).
- The PropS-SH coating with entrapped β-CD–MPT complex exhibited self-healing properties on bronze during exposures to cyclic AR spray at 35 °C.

Supplementary Materials: Supplementary materials are available online at http://www.mdpi.com/2079-6412/9/8/508/s1, Table S1: Parameters related to the titration, Figure S1: Variations of a portion of 1H NMR spectrum (400 MHz) of MPT 2.8 mM during its titration with β-CD 12 mM in D2O, the molar fractions of host are reported in Table S1, Figure S2: Stacked plot and computational analysis of the DOSY experiment of a sample containing MPT alone. MPT signal decays according to the gradients (G) together with the corresponding graphical analysis of the data. The diffusion coefficients are indicated with the letter F, Figure S3. Stacked plot and computational analysis of the DOSY experiment of a sample containing MPT + β-CD. MPT signal decays according to the gradients (G) together with the corresponding graphical analysis of the data. The diffusion coefficients are indicated with the letter F, Figure S4: Stacked plot and computational analysis of the DOSY experiment of a sample containing MPT + β-CD. β-CD signal decays according to the gradients (G) together with the corresponding graphical analysis of the data.

Author Contributions: Conceptualization and Methodology, C.M., G.F., and A.B.; Formal analysis C.M., G.F., and A.B.; Investigation G.F., G.D.C., C.M., F.Z., and A.B.; Writing—Review and Editing C.M., G.F., C.M., F.Z., and A.B.

Funding: This research was carried out in the scope of the B-IMPACT project (project ID 1149) of the M-ERA.NET 2013 call, supported by the national funding organizations of the partners Slovenia (MIZS), Italy (MIUR), and France (RMP).

Acknowledgments: The authors also gratefully acknowledge the University of Ferrara for financial support (FAR funds). Thanks are also given to P. Formaglio for NMR spectroscopic experiments and to T. Bernardi for mass spectrometric experiments.

Conflicts of Interest: The authors declare no conflict of interest.

References

1. Tasić, Ž.Z.; Petrović Mihajlović, M.B.; Radovanović, M.B.; Antonijević, M.M. New trends in corrosion protection of copper. *Chem. Pap.* **2019**, *73*, 2103–2132. [CrossRef]
2. Ashassi-Sorkhabi, H.; Moradi-Alavian, S.; Esrafili, M.D.; Kazempour, A. Hybrid sol-gel coatings based on silanes-amino acids for corrosion protection of AZ91 magnesium alloy: Electrochemical and DFT insights. *Prog. Org. Coat.* **2019**, *131*, 191–202. [CrossRef]
3. Zaferani, S.; Peikari, M.; Zaarei, D.; Danaee, I.; Fakhraei, J.; Mohammadi, M. Using silane films to produce an alternative for chromate conversion coatings. *Corrosion* **2013**, *69*, 372–387. [CrossRef]
4. Bierwagen, G.; Shedlosky, T.J.; Stanek, K. Developing and testing a new generation of protective coatings for outdoor bronze sculpture. *Prog. Org. Coat.* **2003**, *48*, 289–296. [CrossRef]
5. Lamaka, S.V.; Shchukin, D.G.; Andreeva, D.V.; Zheludkevich, M.L.; Möhwald, H.; Ferreira, M.G.S. Sol-gel/polyelectrolyte active corrosion protection system. *Adv. Funct. Mater.* **2008**, *18*, 3137–3147. [CrossRef]
6. Zhang, F.; Ju, P.; Pan, M.; Zhang, D.; Huang, Y.; Li, G.; Li, X. Self-healing mechanisms in smart protective coatings: A review. *Corros. Sci.* **2018**, *144*, 74–88. [CrossRef]
7. Montemor, M.F. Functional and smart coatings for corrosion protection: A review of recent advances. *Surf. Coat. Technol.* **2014**, *258*, 17–37. [CrossRef]
8. Shchukin, D.G.; Zheludkevich, M.; Yasakau, K.; Lamaka, S.; Ferreira, M.G.S.; Möhwald, H.; Shchukin, D. Layer-by-layer assembled nanocontainers for self-healing corrosion protection. *Adv. Mater.* **2006**, *18*, 1672–1678. [CrossRef]
9. Jafari, A.; Hosseini, S.; Jamalizadeh, E. Investigation of smart nanocapsules containing inhibitors for corrosion protection of copper. *Electrochim. Acta* **2010**, *55*, 9004–9009. [CrossRef]
10. Mahmoudian, M.; Nozad, E.; Kochameshki, M.G.; Enayati, M. Preparation and investigation of hybrid self-healing coatings containing linseed oil loaded nanocapsules, potassium ethyl xanthate and benzotriazole on copper surface. *Prog. Org. Coat.* **2018**, *120*, 167–178. [CrossRef]
11. Zheludkevich, M.L.; Poznyak, S.K.; Rodrigues, L.M.; Raps, D.; Hack, T.; Dick, L.F.; Nunes, T.; Ferreira, M.G.S. Active protection coatings with layered double hydroxide nanocontainers of corrosion inhibitor. *Corros. Sci.* **2010**, *52*, 602–611. [CrossRef]

12. Szejtli, J. Introduction and general overview of cyclodextrin chemistry. *Chem. Rev.* **1998**, *98*, 1743–1754. [CrossRef]
13. Hedges, A.R. Industrial applications of cyclodextrins. *Chem. Rev.* **1998**, *98*, 2035–2044. [CrossRef]
14. Uekama, K.; Hirayama, F.; Irie, T. Cyclodextrin drug carrier systems. *Chem. Rev.* **1998**, *98*, 2045–2076. [CrossRef]
15. Del Valle, E.M.M. Cyclodextrins and their uses: A review. *Process Biochem.* **2004**, *39*, 1033–1046. [CrossRef]
16. Li, S.; Purdy, W.C. Cyclodextrins and their applications in analytical chemistry. *Chem. Rev.* **1992**, *92*, 1457–1470. [CrossRef]
17. Takahashi, K. Organic reactions mediated by cyclodextrins. *Chem. Rev.* **1998**, *98*, 2013–2034. [CrossRef]
18. De Souza, T.M.; Cordeiro, R.F.; Viana, G.M.; Aguiar, L.C.; De Senna, L.F.; Malta, L.F.B.; D'Elia, E.; D'Elia, E. Inclusion compounds of dibenzylthiourea with hydroxypropylated-cyclodextrins for corrosion protection of carbon steel in acidic medium. *J. Mol. Struct.* **2016**, *1125*, 331–339. [CrossRef]
19. Casaletto, M.P.; Figà, V.; Privitera, A.; Mazzaglia, A.; Scala, A.; Zagami, R. Sustainable corrosion inhibition of Copper-based alloys by smart β-Cyclodextrin/Benzotriazole complexes. In Proceedings of the 5th European Cyclodextrin Conference, Lisbon, Portugal, 3–6 October 2017.
20. Khramov, A.; Voevodin, N.; Balbyshev, V.; Donley, M. Hybrid organo-ceramic corrosion protection coatings with encapsulated organic corrosion inhibitors. *Thin Solid Films* **2004**, *447*, 549–557. [CrossRef]
21. Khramov, A.N.; Voevodin, N.N.; Balbyshev, V.N.; Mantz, R.A. Sol–gel-derived corrosion-protective coatings with controllable release of incorporated organic corrosion inhibitors. *Thin Solid Films* **2006**, *514*, 174–181. [CrossRef]
22. Amiri, S.; Rahimi, A. Anticorrosion behavior of cyclodextrins/inhibitor nanocapsule-based self-healing coatings. *J. Coat. Technol. Res.* **2016**, *13*, 1095–1102. [CrossRef]
23. Altin, A.; Rohwerder, M.; Erbe, A. Cyclodextrins as carriers for organic corrosion inhibitors in organic coatings. *J. Electrochem. Soc.* **2017**, *164*, C128–C134. [CrossRef]
24. Varvara, S.; Bostan, R.; Bobis, O.; Găină, L.; Popa, F.; Mena, V.; Souto, R.M. Propolis as a green corrosion inhibitor for bronze in weakly acidic solution. *Appl. Surf. Sci.* **2017**, *426*, 1100–1112. [CrossRef]
25. Verma, C.; Ebenso, E.E.; Bahadur, I.; Quraishi, M. An overview on plant extracts as environmental sustainable and green corrosion inhibitors for metals and alloys in aggressive corrosive media. *J. Mol. Liq.* **2018**, *266*, 577–590. [CrossRef]
26. Marzorati, S.; Verotta, L.; Trasatti, S.P. Green corrosion inhibitors from natural sources and biomass wastes. *Molecules* **2019**, *24*, 48. [CrossRef]
27. Zucchi, F.; Trabanelli, G.; Fonsati, M. Tetrazole derivatives as corrosion inhibitors for copper in chloride solutions. *Corros. Sci.* **1996**, *38*, 2019–2029. [CrossRef]
28. Mihit, M.; Salghi, R.; Bazzi, L.; Hammouti, B.; Kertit, S.; El Issami, S.; Addi, E.A. A study of tetrazoles derivatives as corrosion inhibitors of copper in nitric acid. *Pigment Resin Technol.* **2006**, *35*, 151–157. [CrossRef]
29. Balbo, A.; Chiavari, C.; Martini, C.; Monticelli, C. Effectiveness of corrosion inhibitor films for the conservation of bronzes and gilded bronzes. *Corros. Sci.* **2012**, *59*, 204–212. [CrossRef]
30. Wolfe, J.; Grayburn, R. A review of the development and testing of Incralac lacquer. *J. Am. Inst. Conserv.* **2017**, *56*, 225–244. [CrossRef]
31. Aufray, M.; Balbo, A.; Benetti, F.; Bernardi, E.; Bignozzi, M.C.; Chiavari, C.; Esvan, J.; Gartner, N.; Grassi, V.; Josse, C.; et al. Protection of outdoor bronzes by eco-friendly and non-hazardous coatings based on silane and fluoropolymers: Results from the B-IMPACT project. In Proceedings of the Metal 2019, Neuchâtel, Switzerland, 2–6 September 2019; pp. 1–10.
32. Chiavari, C.; Balbo, A.; Bernardi, E.; Martini, C.; Bignozzi, M.; Abbottoni, M.; Monticelli, C.; Bignozzi, M. Protective silane treatment for patinated bronze exposed to simulated natural environments. *Mater. Chem. Phys.* **2013**, *141*, 502–511. [CrossRef]
33. Masi, G.; Balbo, A.; Esvan, J.; Monticelli, C.; Avila, J.; Robbiola, L.; Bernardi, E.; Bignozzi, M.; Asensio, M.; Martini, C.; et al. X-ray photoelectron spectroscopy as a tool to investigate silane-based coatings for the protection of outdoor bronze: The role of alloying elements. *Appl. Surf. Sci.* **2018**, *433*, 468–479. [CrossRef]
34. Chiavari, C.; Bernardi, E.; Balbo, A.; Monticelli, C.; Raffo, S.; Bignozzi, M.; Martini, C.; Bignozzi, M. Atmospheric corrosion of fire-gilded bronze: Corrosion and corrosion protection during accelerated ageing tests. *Corros. Sci.* **2015**, *100*, 435–447. [CrossRef]
35. Wu, D.H.; Chen, A.D.; Johnson, C.S., Jr. An improved diffusion-ordered spectroscopy experiment incorporating bipolar-gradient pulses. *J. Magn. Reson. Ser. A* **1995**, *115*, 260–264. [CrossRef]

36. Fielding, L. Determination of association constants (K_a) from solution NMR data. *Tetrahedron* **2000**, *56*, 6151–6170. [CrossRef]
37. Frassineti, C.; Ghelli, S.; Gans, P.; Sabatini, A.; Moruzzi, M.S.; Vacca, A. Nuclear magnetic resonance as a tool for determining protonation constants of natural polyprotic bases in solution. *Anal. Biochem.* **1995**, *231*, 374–382. [CrossRef]
38. Zucchi, F.; Frignani, A.; Grassi, V.; Trabanelli, G.; DalColle, M. The formation of a protective layer of 3-mercapto-propyl-trimethoxy-silane on copper. *Corros. Sci.* **2007**, *49*, 1570–1583. [CrossRef]
39. Chiavari, C.; Balbo, A.; Bernardi, E.; Martini, C.; Zanotto, F.; Vassura, I.; Bignozzi, M.; Monticelli, C.; Bignozzi, M. Organosilane coatings applied on bronze: Influence of UV radiation and thermal cycles on the protectiveness. *Prog. Org. Coat.* **2015**, *82*, 91–100. [CrossRef]
40. Masi, G.; Josse, C.; Esvan, J.; Chiavari, C.; Bernardi, E.; Martini, C.; Bignozzi, M.C.; Monticelli, C.; Zanotto, F.; Balbo, A.; et al. Evaluation of the protectiveness of an organosilane coating on patinated Cu-Si-Mn bronze for contemporary art. *Prog. Org. Coat.* **2019**, *127*, 286–299. [CrossRef]
41. Stern, M.; Geary, A.L. Electrochemical polarization I. A theoretical analysis of the shape of polarization curves. *J. Electrochem. Soc.* **1957**, *104*, 56–63. [CrossRef]
42. Pessine, F.B.T.; Calderini, A.; Alexandrino, G.L. Review: Cyclodextrin inclusion complexes probed by NMR techniques. *Magn. Reson. Spectr.* **2012**, *2012*, 237–264.
43. Kfoury, M.; Auezova, L.; Greige-Gerges, H.; Ruellan, S.; Fourmentin, S. Cyclodextrin, an efficient tool for trans-anethole encapsulation: Chromatographic, spectroscopic, thermal and structural studies. *Food Chem.* **2014**, *164*, 454–461. [CrossRef]
44. Schneider, H.-J.; Hacket, F.; Rüdiger, V.; Ikeda, H. NMR studies of cyclodextrins and cyclodextrin complexes. *Chem. Rev.* **1998**, *98*, 1755–1786. [CrossRef]
45. Bothner-By, A.A.; Stephens, R.L.; Lee, J.; Warren, C.D.; Jeanloz, R.W. ChemInform Abstract: Structure determination of a tetrasaccharide: transient nuclear overhauser effects in the rotating frame. *Chem. Inf.* **1984**, *15*, 811–813. [CrossRef]
46. Holm, R.; Østergaard, J.; Schőnbeck, C.; Jensen, H.; Shi, W.; Peters, G.H.; Westh, P. Determination of stability constants of tauro- and glycoconjugated bile salts with the negatively charged sulfobutylether-β-cyclodextrin: Comparison of affinity capillary electrophoresis and isothermal titration calorimetry and thermodynamic analysisof the interaction. *J. Incl. Phenom. Macrocycl. Chem.* **2014**, *78*, 185–194. [CrossRef]
47. Cameron, K.S.; Fielding, L. NMR diffusion coefficient study of steroid–cyclodextrin inclusion complexes. *Magn. Reson. Chem.* **2002**, *40*, S106–S109. [CrossRef]
48. Rymdén, R.; Carlfors, J.; Stilbs, P. Substrate binding to cyclodextrins in aqueous solution: A multicomponent self-diffusion study. *J. Incl. Phenom. Macrocycl. Chem.* **1983**, *1*, 159–167.
49. Mayzel, O.; Cohen, Y. Diffusion coefficients of macrocyclic complexes using the PGSE NMR technique: Determination of association constants. *J. Chem. Soc. Chem. Commun.* **1994**, *16*, 1901–1902. [CrossRef]
50. Wimmer, R.; Aachmann, F.L.; Larsen, K.L.; Petersen, S.B. NMR diffusion as a novel tool for measuring the association constant between cyclodextrin and guest molecules. *Carbohydr. Res.* **2002**, *337*, 841–849. [CrossRef]
51. Cameron, K.S.; Fielding, L. NMR diffusion spectroscopy as a measure of host–guest complex association constants and as a probe of complex size. *J. Org. Chem.* **2001**, *66*, 6891–6895. [CrossRef]
52. Šmejkalová, D.; Piccolo, A. Host-guest interactions between 2,4-dichlorophenol and humic substances as evaluated by ^{1}H NMR relaxation and diffusion ordered spectroscopy. *Environ. Sci. Technol.* **2008**, *42*, 8440–8445. [CrossRef]
53. Ye, X.; Xin, X.; Zhu, J.; Xue, Z. Coordination compound films of 1-phenyl-5-mercaptotetrazole on copper surface. *Appl. Surf. Sci.* **1998**, *135*, 307–317. [CrossRef]
54. Monticelli, C.; Balbo, A.; Esvan, J.; Chiavari, C.; Martini, C.; Zanotto, F.; Marvelli, L.; Robbiola, L. Evaluation of 2-(salicylideneimino) thiophenol and other Schiff bases as bronze corrosion inhibitors by electrochemical techniques and surface analysis. *Corros. Sci.* **2019**, *148*, 144–158. [CrossRef]

coatings

Article

Multifunctional Protective PVC-ZnO Nanocomposite Coatings Deposited on Aluminum Alloys by Electrospinning

Alvaro Iribarren [1], Pedro J. Rivero [1,2,*], Carlos Berlanga [1,2], Silvia Larumbe [3], Adrian Miguel [3], Jose F. Palacio [3] and Rafael Rodriguez [1,2]

1 Materials Engineering Laboratory, Department of Engineering, Public University of Navarre, Campus Arrosadía S/N, 31006 Pamplona, Spain; iribarren.111228@e.unavarra.es (A.I.); carlos.berlanga@unavarra.es (C.B.); rafael.rodriguez@unavarra.es (R.R.)

2 Institute for Advanced Materials (InaMat), Public University of Navarre, Campus Arrosadía S/N, 31006 Pamplona, Spain

3 Centre of Advanced Surface Engineering, AIN, 31191 Cordovilla, Spain; SLarumbe@ain.es (S.L.); AMiguel@ain.es (A.M.); jfpalacio@ain.es (J.F.P.)

* Correspondence: pedrojose.rivero@unavarra.es; Tel.: +34-948-168-961

Received: 8 March 2019; Accepted: 25 March 2019; Published: 27 March 2019

Abstract: This paper reports the use of the electrospinning technique for the synthesis of nanocomposite micro/nanofibers by combining a polymeric precursor with hydrophobic behavior like polyvinyl chloride (PVC) with nanoparticles of a corrosion inhibitor like ZnO. These electrospun fibers were deposited on substrates of the aluminum alloy 6061T6 until forming a coating around 100 μm. The effect of varying the different electrospinning deposition parameters (mostly applied voltage and flow-rate) was exhaustively analyzed in order to optimize the coating properties. Several microscopy and analysis techniques have been employed, including optical microscopy (OM), field emission scanning electron microscopy (FE-SEM), atomic force microscopy (AFM), thermogravimetric analysis (TGA), and differential scanning calorimetry (DSC). Water contact angle (WCA) measurements have been carried out in order to corroborate the coating hydrophobicity. Finally, their corrosion behavior has been evaluated by electrochemical tests (Tafel curves and pitting potential measurements), showing a relevant improvement in the resultant corrosion resistance of the coated aluminum alloys.

Keywords: electrospinning; corrosion; superhydrophobic; PVC-ZnO; nanocomposite coating

1. Introduction

Aluminum alloys are widely used in many industrial areas such as automotive, aerospace, and construction, due to their physical and mechanical properties: Low density, good mechanical properties at low temperatures, good corrosion resistance, and excellent thermal and electrical conductivities [1]. While aluminum is a very reactive metal with a high affinity for oxygen, the metal shows a high corrosion resistance in most environments. This is due to the thin layer of aluminum oxide (Al_2O_3) that is formed on the surface of the metal. This passive layer, of around 5 to 10 nanometers thick, shows an inert and protective character and acts as a barrier between the metal and the surrounding medium. However, in spite of this passive layer, Al and its alloys can still suffer corrosion when they are exposed to aggressive corrosive media like those containing chloride ions [2]. In order to extend the range of applications of these alloys, it is necessary to extend their corrosion resistance by means of surface treatments or coatings with the aim to reduce the economic impact of corrosion [3].

A widely used anti-corrosion practice for metals is the implementation of barrier coatings, which consists of covering the metallic surface with a protective film that acts as a barrier and avoids the direct interaction between the metal and the environment. Traditional coatings containing Cr and Pb have been banned in many countries because they are a threat for human health. This is why the development of ecofriendly polymeric coatings has gained special interest nowadays [4]. In particular, a technique that is increasing in popularity in the last few years is electrospinning due to its relative ease of use and because it allows the deposition ultrafine polymeric fibers over a metallic surface that can act as an effective barrier coating and reduce the corrosion rate of the metal.

Recently, superhydrophobic (SH) coatings with water contact angle (WCA) higher than 150° have gained special interest due to their excellent water repellency properties. These SH coatings can be used to improve the corrosion resistance of Al and Al alloys as they minimize the contact area between the metallic surface and the corrosive electrolyte. SH surfaces can be produced by increasing the roughness of the surface for the target material [5–7]. Electrospinning is a powerful technique for developing SH coatings because the electrospun micro/nanofibers form a film that presents a surface roughness high enough to provide hydrophobic or even superhydrophobic behavior [8,9]. In addition, electrospun coatings can be obtained in a short period of time by using only one step and the diameter thickness of the resultant fibers can be perfectly controlled from nanometric to micrometric scale by changing the fabrication parameters (applied voltage, flow rate, and distance collector) as well as the resultant viscosity of the precursor, respectively

Corrosion resistance can be further enhanced by the addition of a corrosion inhibitor, either organic or inorganic. Many researchers have focused on blending a polymer with inorganic nanoparticles, such as Al_2O_3, ZrO_2, SiO_2, TiO_2, Fe_3O_4, CdS, and ZnO, among others [10–19]. The presence of these nanoparticles between the interstices of the fibers can increase the surface roughness and the air entrapment, improving the hydrophobic character of the coating and thus reducing the corrosion rate.

In this work, polyvinyl chloride (PVC) was the polymer chosen for the development of anticorrosion electrospun coatings because it is a polymer with hydrophobic behavior by nature, and zinc oxide (ZnO) nanoparticles were used as the corrosion inhibitor because the corrosion behavior of electrospun PVC-ZnO, to the best of our knowledge, has not been addressed previously. PVC-ZnO nanofibers were successfully deposited onto aluminum 6061T6 substrates by electrospinning and the corrosion behavior of the nanocomposite film was exhaustively analyzed.

2. Experimental Procedure

2.1. Reagents and Materials

Polyvinyl chloride (PVC, $(C_2H_3Cl)_n$, M_w = 80,000 g/mol), dimethylformamide (DMF) and tetrahydrofuran (THF) were purchased from Sigma-Aldrich (Saint Louis, MO, USA). Zinc oxide (ZnO) nanoparticles in colloidal dispersion (50% in H_2O) with average particle size of 50 nm were purchased from Alfa Aesar (Haverhill, MA, USA). All reagents used without further purification. The coatings have been performed onto aluminum substrates of the alloy AA6061T6.

2.2. Deposition of PVC-ZnO Nanocomposite Coatings

PVC was dissolved in a 1:1 solvents mixture of THF and DMF, thus obtaining a homogeneous PVC solution with polymer concentration 15 wt %, respectively. The viscosity of the solvents was 0.92 mPa·s for DMF and 0.55 mPa·s for THF, respectively. An important aspect to remark is that the final solution viscosity depends on the polymer concentration and the volume ratio between the solvents [20,21]. The solution was prepared under vigorous stirring (600 rpm) for 12 h at room temperature by using a magnetic stirrer. Separately, 0.64 g of the zinc oxide colloidal dispersion was added to a mixture of 1 mL of THF and 1 mL of DMF (same 1:1 solvent volume ratio as the polymer solution). This ZnO mixture was produced and dispersed applying stirring (200 rpm) for 12 h at room temperature. Afterwards, the dispersed ZnO mixture was added to the PVC solution, and then vigorous stirring (600 rpm) was applied for 2 h in order to allow for an adequate mixing.

The PVC-ZnO solution was then electrospun using ND-ES 11/7 Lab Electrospinning Unit (Nadetech Innovations S.L., Navarre, Spain). A horizontal configuration with fixed collector was used, and the capillary tip was located 15 cm away from the aluminum electrode that was used as the collector. A 20-gauge needle with an inner diameter specific of 0.6 mm has been used for all the experiments. Electrospinning experiments were conducted at various pump speeds and direct current (DC) voltages in order to optimize the values of flow rate and applied voltage that allow for the fabrication of very thin fibers with maximum water repellency. Finally, in all the experiments the electrospinning process was conducted at room temperature (20 °C) at 40% relative humidity and the deposition time was fixed for a period of time of 10 min, respectively.

2.3. Characterization Techniques

Optical microscopy (OM, BX60M, Olympus, Tokyo, Japan) was used for a first observation of fiber morphology and for estimating the average fiber diameter. Field emission scanning electron microscopy (FE-SEM, S4800, Hitachi, Tokyo, Japan) and atomic force microscopy (AFM, Veeco Innova AFM, Veeco Instruments, Plainview, NY, USA) were employed for a more exhaustive analysis of the fiber size, morphology and surface roughness.

Water contact angle (WCA) measurements were conducted on a CAM 100 contact angle goniometer (CAM 100, KSV Instruments, Burlington, VT, USA) using distilled water. The static water contact angle was measured 5 times at different sites and the representative samples were chosen to present the result.

The composition of the films was controlled by thermal gravimetric analysis (TGA, Q500, TA Instruments, New Castle, DE, USA), and differential scanning calorimetry (DSC, DSC25, TA Instruments, New Castle, DE, USA) was used for studying the thermal properties of the coatings. DSC tests were performed starting from an equilibrium temperature of −80 °C and raising 10 °C per minute up to a final temperature of 200 °C. In addition, the adhesion of the coating has been determined by applying a pressure-sensitive adhesive tape onto a grid of small squares previously formed. The fraction of coating removed from the grid when the adhesive tape is removed is compared with standard ratings.

Electrochemical measurements including Tafel polarization curves and pitting corrosion tests were carried out on an Autolab Potentiostat/Galvanostat PGSTAT302N (Metrohm, Herisau, Switzerland). All corrosion tests were performed at room temperature in 3.5 wt % NaCl aqueous solutions, using a conventional three electrode cell consisting of a working electrode (bare or coated Al sample), a silver chloride Ag-AgCl reference electrode and a platinum counter electrode. Before conducting all the experiments, the samples were immersed in the 3.5 wt % NaCl electrolyte for 30 min to make sure that the system is in steady state. Tafel polarization measurements were obtained by scanning the electrode potential automatically from −150 to +150 mV at a scan rate of 1.5 $mV \cdot s^{-1}$. The corrosion protection efficiency from Tafel polarization curves was calculated by the following formula [22,23]:

$$\eta = \frac{i_{corr} - i_{corr}(C)}{i_{corr}} \times 100\% \quad (1)$$

In this equation, i_{corr} and $i_{corr}(C)$ correspond to the corrosion current densities of bare aluminum and coated aluminum, respectively.

For the pitting corrosion tests the measurements were obtained by scanning the electrode potential from 0 to +2 V at a scan rate of 0.15 $mV \cdot s^{-1}$, and reversing the scan direction once the current reaches a cutoff value of 2.5 mA.

3. Results and Discussion

3.1. Morphologies and Wettability

The first step was the optimization of the electrospinning parameters, mainly applied voltage (E) and flow rate. These two are the processing parameters that have a bigger impact on the final morphology of the resultant fibers. Several samples of PVC were electrospun under different electrospinning conditions, varying these two parameters. Firstly, the lower and upper limit values of the electric potential that can be applied for a successful deposition need to be determined. Voltage values of 6 kV and lower showed to be too low and no fibers were deposited. There is also an upper limit for *E* above which the Taylor cone is no longer formed and the fiber jet is ejected from within the needle, which is associated with an increase in fiber defects [24]. The Taylor cone was observed to be formed in the range of 8–14 kV. These values were chosen to be the lower and upper limits of the applied voltage for the set of experiments of the parameter optimization process. In Figure 1a it can be seen that fiber diameter of PVC electrospun fibers decreases when increasing the applied voltage, which goes in accordance with the results obtained by Lee et al. [25]. This may be explained by an increase in the electrostatic repulsive force on the charged jet narrowing the fiber diameter. In addition, Figure 1b shows that the resultant fiber diameter gets bigger when increasing the flow rate, as expected, which is in concordance with literature [24,26]. Other aspects to remark is that, despite the variation of fiber diameter, no remarkable differences in morphology were observed when applying different voltages and flow rates.

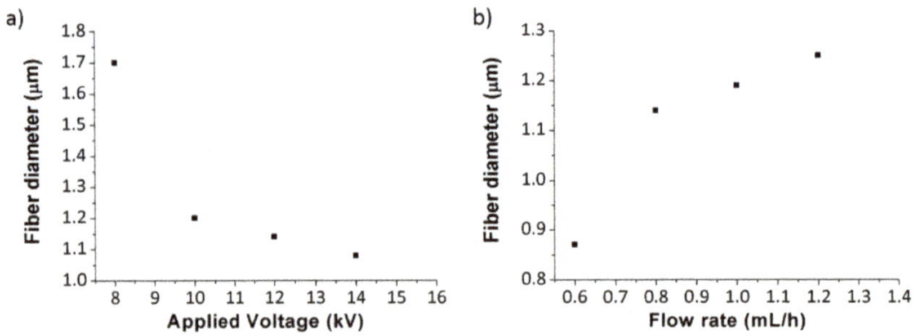

Figure 1. (**a**) Evolution of the resultant fiber diameter as a function of the applied voltage with a specific fixed flow rate of 0.8 mL·h^{-1}; and (**b**) evolution of the fiber diameter as a function of the flow rate with a specific fixed applied voltage of 12 kV.

Once it has been evaluated the effect of the experimental parameters on the resultant fiber diameter, the PVC-ZnO fibers in this work have been electrospun under 14 kV and a flow rate of 0.6 mL·h^{-1} because there are scientific studies which indicate that there is an increase in the hydrophobicity caused by a reduction in diameter among bead-free fibers [27]. Figure 2 shows SEM micrographs revealing the morphology of PVC-ZnO nanocomposite fibers with different image amplifications. It can be clearly seen that the resultant electrospun nanofibers present relatively uniform fibers whose average diameter was measured to be around 720 nm.

In order to corroborate the presence of zinc oxide into the electrospun nanofibers, the composition of the nanocomposite fibers (PVC-ZnO) has also been analyzed by using a thermal gravimetric analysis (TGA), as it can be appreciated in Figure 3. First of all, the TGA thermogram obtained from blank films (only PVC without ZnO) showed two consecutive weight losses (solid line) around 345 and 490 °C, whereas no further weight loss has been observed for higher values of temperature. However, the sample composed of PVC-ZnO nanoparticles showed a third weight loss around 530 °C (dash-dotted line), which corresponds to the amount of ZnO nanoparticles (1.34% in weight) into the electrospun PVC fibers.

The surface wettability of the PVC-ZnO nanocomposite coating was assessed performing water contact angle (WCA) measurements. In Figure 4, representative optical images of water droplets deposited onto both PVC and PVC-ZnO electrospun fiber mats are shown. It is worthy to mention that similar contact angle values in the range of 145°–155° have been measured over both samples, so no significant differences in the wettability have been observed between the samples composed of ZnO nanoparticles in comparison with only PVC samples.

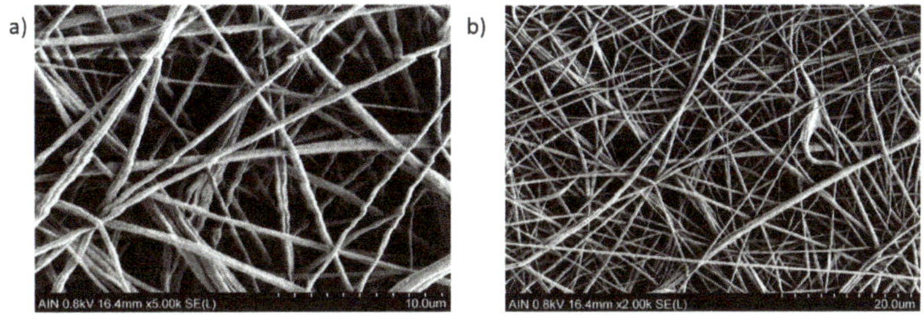

Figure 2. SEM images of the electrospun PVC-ZnO nanocomposite fiber at different scale bar of 10 μm (**a**) and 20 μm (**b**), respectively.

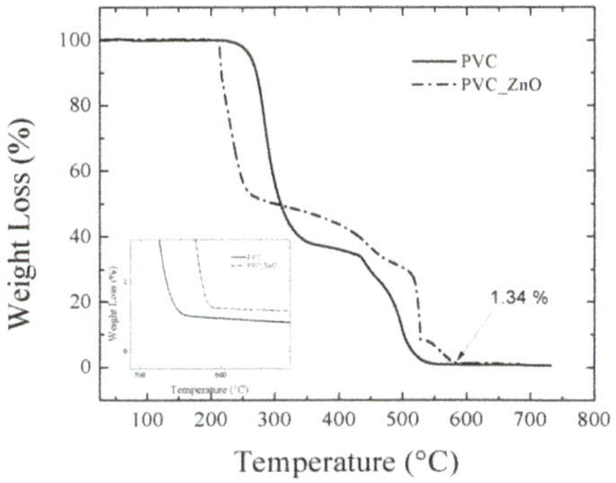

Figure 3. Thermal gravimetric analysis (TGA) of both only PVC fibers (solid line) and PVC fibers containing ZnO nanoparticles (dash-dotted line).

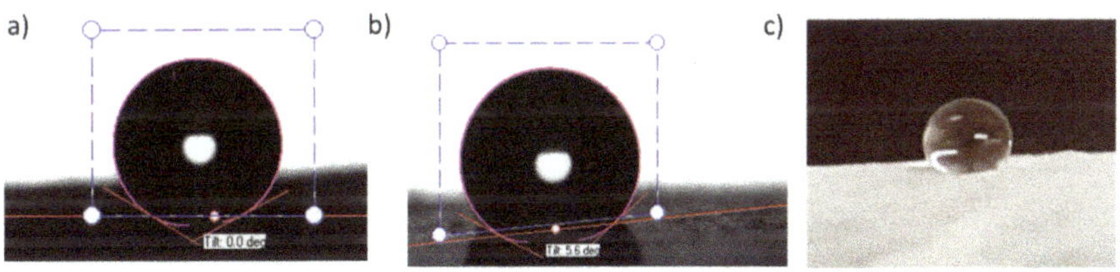

Figure 4. Optical images of the water contact angle values on the PVC sample with a value of 151.73° (**a**), PVC-ZnO sample 146.39° (**b**) image of a water drop deposited on a surface coated with electrospun fibers where the high water repellency can be observed and (**c**) aspect of the droplet onto electrospun coating.

Low adhesive forces were noticed between the PVC-ZnO nanocomposite substrate and water droplets, which may be attributed to the high surface roughness of the fibrous mat. Initially, the wetting of rough surfaces was explained by the Wenzel [28], or the Cassie-Baxter or models [29]. This later model states that surface superhydrophobicity is caused by the air entrapment underneath the liquid inside the grooves. In the Cassie-Baxter state, the contact angle of a water droplet on a hydrophobic surface (θ_c) is related to the contact angle on a smooth surface (θ) by the following equation:

$$\cos(\theta_c) = f_1 \cos(\theta) - f_2 \quad (2)$$

In this equation f_1 and f_2 are proportions of solid surface and air in contact with liquid, respectively. In addition, $f_1 + f_2 = 1$.

More recently, both, the Wenzel and the Cassie-Baxter models have been revised and, in some extent, highly criticized [30,31]. In most of published papers about wettability of electrospun coatings, authors employ the Cassie-Baxter model to explain the main features [32], although mixed states and transitions between both phases have been reported [33,34]. In particular, the effects of fiber diameter, fiber density and liquid surface tension are taken in account. In general, the air trapped between fibers keeps the wetting behavior in the Cassie-Baxter regime, but transitions to the Wenzel state are reported for liquids with surface tension below 58 $mN \cdot m^{-1}$, which is not the case in water [35].

Assuming a fair validity of the Cassie-Baxter model, as the contact angle value of the PVC-ZnO nanocomposite coating was found to be $\theta_c = 147°$, their equation provides a value for f_2 of, approximately, 0.84. This means that about 84% of the water drop is in contact with air, and only 16% is in contact with the solid surface. Such hydrophobic behavior is desired in an anticorrosion coating because, if there is small contact area between the electrolyte and the metal, a less amount of aggressive ions will attack the metallic surface.

Atomic force microscopy was used in order to examine the morphology and to determine the surface roughness of the prepared electrospun nanocomposite coatings. In Figure 5 is shown in 2-dimensional (2D) and 3-dimensional (3D) AFM images the PVC-ZnO nanocomposite coating electrospun at 14 kV and a flow rate of 0.6 $mL \cdot h^{-1}$. In addition, the surface roughness was investigated with AFM by measuring 3 evaluation lines which are presented in Figure 6. It was observed that the average roughness R_a of the PVC-ZnO coating is 427.5 nm.

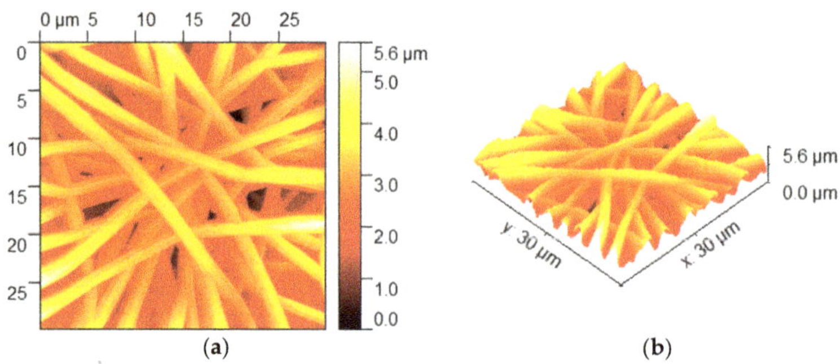

Figure 5. AFM images of PVC-ZnO sample in 2-D (**a**) and 3-D (**b**). Image dimensions: 30 μm × 30 μm.

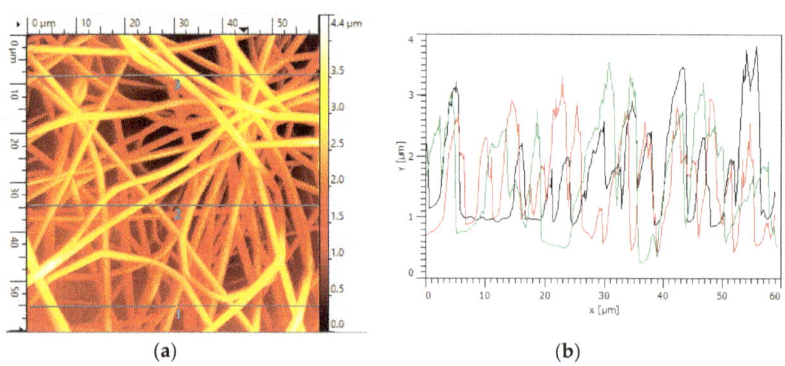

Figure 6. (**a**): 2D AFM image of the PVC-ZnO sample. Image dimensions: 60 μm × 60 μm; (**b**): roughness profiles corresponding to the three evaluation lines indicated (**a**).

3.2. *Effect of Heat Treatment on Hydrophobicity*

In this work the behaviour of the coatings after a thermal treatment was studied. The main limitation found for the electrospun coatings is their low adherence to the substrate. This can be an obstacle for the implementation of this technique in industrial applications. One effective way to increase the adherence of the fiber mat onto the substrate is to perform a thermal treatment. Homaeigohar et al. found that heating of electrospun polyethersulfone fibers (PES) improves adhesion between the electrospun fibers and the underlying substrate [36]. Furthermore, several studies have shown that heat treatment is able to increase the mechanical strength of the electrospun mat. The reason for this may be that heat treatment encourages fusion at the contact points between fibers, providing a strengthening effect [37]. In this work, a crosshatch test has been also performed in order to corroborate this increase in the resultant adhesion onto the underlying substrate after thermal treatment. Firstly, on one hand, when the cutting tool was applied onto the non-heat treated coating, the electrospun fiber mat was heavily damaged, being impossible to perform the crosshatch test. However, on the other hand, when the electrospun coating has been thermally treated, the resultant fiber film has been not been destroyed when performing the crosshatch test. In addition, the electrospun coating still showed a low adhesion onto the aluminum substrate because more than 65% of the area was removed when pulling off the tape.

Heat treatment is usually carried out between the glass transition temperature (T_g) and the melting temperature (T_m) of the material [38]. The application of a heat treatment to the PVC electrospun coating was performed with two different goals: (1) Increasing the adhesion between the fibers and the underlying substrate and (2) checking the hypothesis that heating the sample above T_g would improve the ZnO nanoparticles distribution be to an improvement in tween the fibers due to the movement and vibrations of the polymer chains, thus leading the corrosion resistance of the coating. The glass transition temperature of PVC was determined using differential scanning calorimetry (DSC). Figure 7 shows the DSC plots of the PVC electrospun samples with and without ZnO nanoparticles inclusions. Results showed that the addition of ZnO nanoparticles produced an increment in the amount of heat required to increase the temperature of the electrospun samples, but the shape of the graphs (heat flow versus temperature) remained almost equal. This means that the presence of ZnO nanoparticles in between the fiber mat does not produce a relevant change in the thermal behaviour of PVC fibers. The glass transition temperature of PVC and PVC-ZnO was found to be 82.27 and 82.36 °C respectively. An aspect to remark is that glass transitions can sometimes be confused with an endothermic transition due to the occurrence of molecular relaxation processes in the material. In Figure 7, DSC curves exhibited these molecular relaxation processes related to the glass transition.

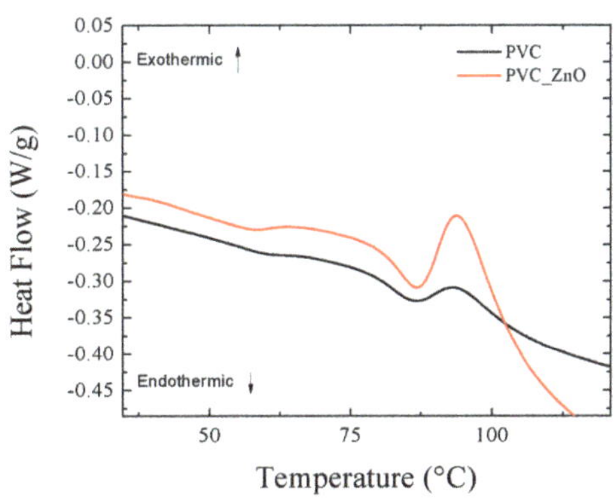

Figure 7. Differential Scanning Calorimetry analysis of both PVC fibers with no corrosion inhibitor (black line) and PVC fibers containing ZnO nanoparticles (red line) with the assignment of the exothermic and endothermic transitions, respectively.

The effects of heat treatment on the WCA values of electrospun PVC fibers at 60, 70, 80, 90, 100, 110 and 120 °C were studied. All heat treatments were performed in an industrial oven during 1 h. In Table 1, the WCA values of PVC fibers obtained at the different temperatures is presented.

Table 1. Effect of heat treatment on the water contact angles values of PVC fiber films.

	Contact Angle Values at Different Temperatures						
Temperature (°C)	60	70	80	90	100	110	120
WCA	147	149	147	139	138	135	120

The contact angle values of PVC fibers started to decrease after surpassing the glass transition temperature of the polymer (i.e., 80 °C) reaching a minimum WCA value of 120° after the final heat treatment at 120 °C. Representative images of this behaviour can be observed in Figure 8. The experimental results are in accordance with the ones obtained by Asmatulu et al. [39]. The change in the contact angle values may be related to the rearrangement of the fiber structures in the film at the glass transition temperature. Below T_g polyvinyl chloride is a glassy solid and the polymer chains are rigid and cannot move. Above T_g the polymer chains start moving and vibrating, and the film becomes softer and rubbery, which may produce this decrease in the water contact angle values.

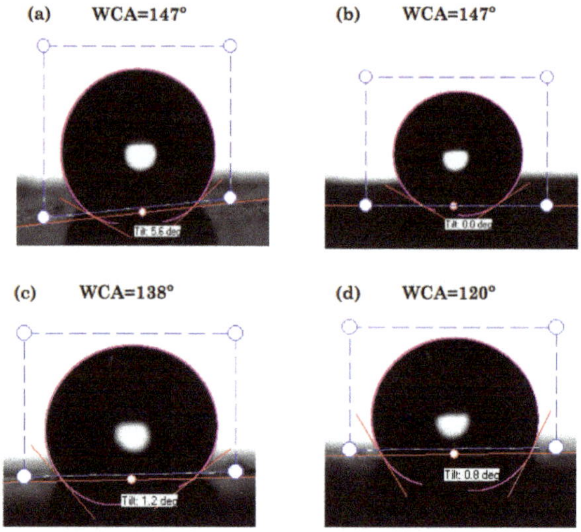

Figure 8. Evolution of the contact angle of the PVC film with increasing temperature. (**a**) heated to 60 °C, (**b**) heated to 80 °C, (**c**) heated to 100 °C and (**d**) heated to 120 °C. The reduction of the WCA when heating over the glass transition temperature of PVC (i.e., 80 °C) can be clearly seen.

3.3. Anticorrosion Performance of PVC-ZnO Nanostructures

The main objective of this work was the electrospinning of polymer coatings with the aim of improving the corrosion resistance of aluminum substrates. In order to check if the electrospun PVC and PVC-ZnO nanocomposite films successfully enhance this corrosion resistance, Tafel polarization and pitting corrosion tests were held. Firstly, corrosion tests were performed over the bare aluminum 6061T6 without any coating. Results are shown in Figure 9. The Tafel analysis showed that the corrosion current density (i_{corr}) for the bare aluminum substrate was 1.107 $\mu A \cdot cm^{-2}$ at a corrosion potential (E_{corr}) of −0.856 V, and the cathodic (β_c) and anodic (β_a) Tafel constants were found to be 150 and 56 mV/decade, respectively. The pitting corrosion curve provides information about the bare aluminum behaviour against localized (pitting) corrosion. Scully et al. defined the pitting potential (E_p)

as the potential above which the rising current permanently exceeded 10 μA [40]. Therefore following that definition the pitting potential of the bare aluminum sample was found to be −0.656 V.

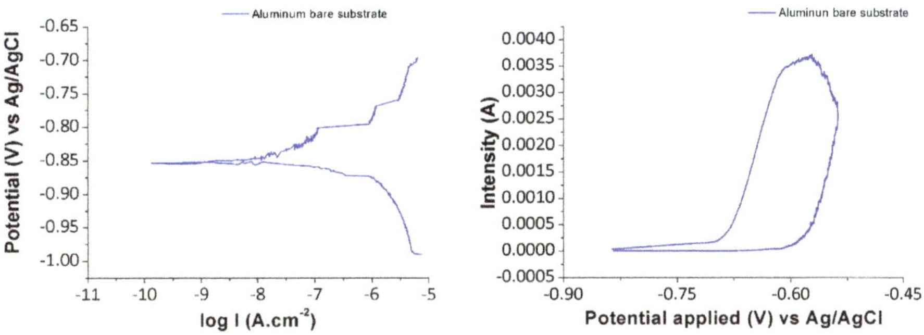

(**a**) (**b**)

Figure 9. Tafel plot (**a**) and pitting corrosion curve (**b**) for the aluminum bare substrate in 3.5 wt % NaCl aqueous solution.

After corrosion tests were successfully conducted over the reference bare Al sample, the corrosion behaviour of aluminum coated with the electrospun polymer films was studied in order to see if there was an improvement in the corrosion resistance of the metallic substrate with the use of these nanocomposite coatings. For testing the corrosion resistance of electrospun PVC films in the presence and absence of ZnO nanoparticles inclusions, corrosion tests were performed over samples after being heated up to the glass transition temperature (T_g) due to the improved adherence of the fiber mat with the metallic substrate, and also over samples heated 20 °C over this temperature (T_g + 20) in order to check the hypothesis that a better distribution of ZnO nanoparticles caused by the movement of the polymer chains above T_g could compensate the contact angle decrease and improve the corrosion resistance of the coating. The Tafel plots are displayed in Figure 10 and results are summarized in Table 2, respectively.

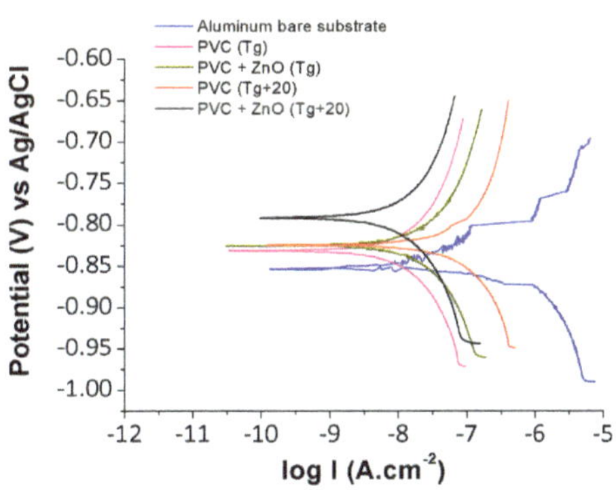

Figure 10. Tafel plots corresponding to the Al bare substrate and the four Al samples coated with electrospun PVC and PVC-ZnO nanocomposite fibers, tested in 3.5 wt % NaCl aqueous solution.

The results show that all electrospun coatings reduced the corrosion current density of aluminum in two orders of magnitude. This corrosion resistance was lowered significantly from 1.107 μA·cm^{-2} in the case of bare Al to a minimum of 0.009 μA·cm^{-2} in case of Al/PVC-ZnO after being heat treated to 100 °C (T_g + 20). The corrosion potential was displaced towards less negative values, attaining a maximum of −0.793 V in the case Al/PVC-ZnO (T_g + 20) again. The protection efficiency (η) in the case of pure PVC after being heat-treated to 80 °C (T_g) was found to be 99.01%. When the pure PVC coating was heated to 100 °C the efficiency was reduced to 97.38%, which can be explained by the decrease in the water contact angle when surpassing the glass transition temperature of the polymer. The composite coating of PVC-ZnO presented lower efficiency than the other samples when heated to T_g, but it was found that when this composite coating was heated 20° higher than T_g the protection efficiency reached a maximum of 99.19%, which may be related to a better distribution of zinc oxide nanoparticles.

Table 2. Tafel analysis for uncoated Al substrate, pure PVC and PVC-ZnO nanocomposite coatings after being heat-treated, tested in 3.5 wt % NaCl aqueous solution.

Sample	βa (mV/dec)	βc (mV/dec)	i_{corr} (μA/cm²)	E_{corr} (V)	η (%)
Al substrate	56	150	1.107	−0.856	–
PVC (T_g)	155	143	0.011	−0.833	99.01
PVC (T_g + 20)	197	172	0.029	−0.827	97.38
PVC-ZnO (T_g)	140	96	0.054	−0.826	95.12
PVC-ZnO (T_g + 20)	175	159	0.009	−0.793	99.19

As the composite coating of PVC-ZnO heat-treated to 100 °C presented the best anti-corrosion properties in the Tafel polarization tests, its behaviour against pitting corrosion was also tested. Figure 11 shows that the pitting corrosion resistance was clearly enhanced by the nanocomposite electrospun coating. Its pitting potential (i.e., the potential above which the rising current permanently exceeded 10 μA [40]) was found to be −0.341 V, which is closer to positive values than the pitting potential of bare Al which was −0.656 V. This means that higher electric potential is needed in order for pitting corrosion to appear in the composite coating. Furthermore, once pitting starts it grows with a slower rate. The current cutoff of 2.5 mA was reached under a potential of 0.015 V in the case of the PVC-ZnO coating, while it was reached at −0.538 V in the case of the bare Al substrate. The pitting current increase rate can be estimated as the slope of the polarization curve between the pitting potential (E_p) and the electric potential at the 2.5 mA current cutoff point (E_{cutoff}). This can be approximated using the following equation:

$$i_{corr} = \frac{2.5 - 10 \times 10^{-3}}{E_{cutoff} - E_p} \text{ (mA/V)} \quad (3)$$

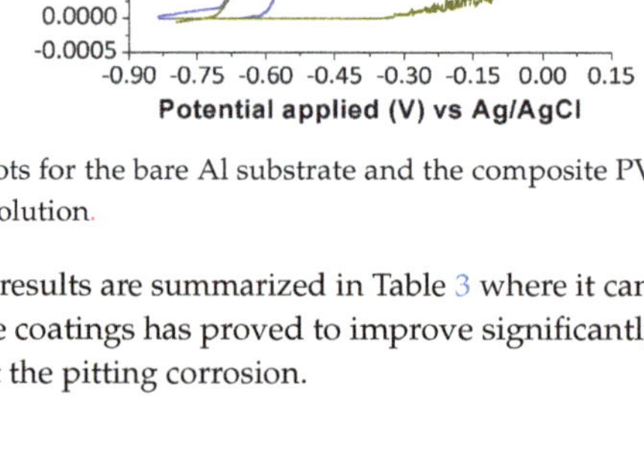

Figure 11. Pitting test plots for the bare Al substrate and the composite PVC-ZnO coating, tested in 3.5 wt % NaCl aqueous solution.

Finally, the pitting test results are summarized in Table 3 where it can be clearly observed that the electrospun nanocomposite coatings has proved to improve significantly the corrosion resistance of the aluminum alloy against the pitting corrosion.

Table 3. Pitting test results for the bare Al substrate and the composite PVC-ZnO coating, tested in 3.5 wt % NaCl aqueous solution.

Sample	E_p (V)	E_{cutoff} (V)	i_{corr} (mA/V)
Al substrate	−0.656	−0.538	21.10
PVC-ZnO (T_g + 20)	−0.341	0.015	6.99

4. Conclusions

In summary, it has been demonstrated that electrospun nanocomposite coatings of PVC improve the corrosion resistance of the aluminum alloy 6061T6 due to the high hydrophobic character (even superhydrophobic) of the resulting surfaces, which may be explained by the roughness and fibrous topography of these coatings, which seems to produce a water-coating contact corresponding to the Cassie-Baxter state.

Nanocomposite films of PVC containing ZnO nanoparticles were successfully prepared using one-step electrospinning technique and the corrosion electrochemical tests revealed that heating the nanocomposite PVC-ZnO structures 20° higher than T_g of the polymer allows obtaining an excellent and better anti-corrosion behaviour. This may be explained by a better distribution of the ZnO nanoparticles due to the movement of the polymer chains above T_g, which compensates for the decrease in the hydrophobic behaviour of the fibers at such high temperatures. Finally, Tafel polarization tests have shown that the corrosion current density could be reduced in two orders of magnitude with the use of these electrospun coatings and pitting corrosion tests also demonstrated that the nanocomposite surfaces enhance the resistance of aluminum against localized corrosion.

Finally, it is worth to notice that real life conditions can be far different from those of static laboratory tests. In particular water condensation or the impact of water drops can lead to a Wenzel-like state that can limit the employment of these coatings in automotive or aeronautical applications. The dynamic test should be necessary to guarantee the good behavior of these coatings in real life applications.

Author Contributions: Conceptualization, A.I. and P.J.R.; Methodology, A.I. and P.J.R.; Validation, A.I., P.J.R., and R.R.; Formal Analysis, A.I., P.J.R., C.B and J.F.P.; Investigation, A.I., P.J.R., S.L., A.M., J.F.P. and R.R.; Writing—Original Draft Preparation, P.J.R. and A.I.; Writing—Review and Editing, A.I., P.J.R. and R.R.; Supervision, P.J.R., C.B. and R.R.

Funding: This research was funded by the Spanish Economy and Competitiveness Ministry, FEDER (Project TRA2013-48603-C4-1-R-HELADA) and by the Public University of Navarra collaboration research grant PRO-UPNA 18 (6107).

Acknowledgments: The authors would like to express their gratitude to Nadetech Inc. for the tune-up of the robot used for the deposition of the nanocoatings. Finally the authors would like to express their grateful acknowledgement for the support received from the Asociación de la Industria Navarra (AIN).

Conflicts of Interest: The authors declare no conflict of interest.

References

1. Abdel-Gaber, A.M.; Abd-El-Nabey, B.A.; Sidahmed, I.M.; El-Zayady, A.M.; Saadawy, M. Kinetics and thermodynamics of aluminium dissolution in 1.0 M sulphuric acid containing chloride ions. *Mater. Chem. Phys.* **2006**, *98*, 291–297. [CrossRef]
2. Badawy, W.A.; Al-Kharafi, F.M.; El-Azab, A.S. Electrochemical behaviour and corrosion inhibition of Al, Al-6061 and Al–Cu in neutral aqueous solutions. *Corros. Sci.* **1999**, *41*, 709–727. [CrossRef]
3. NACE International. Corrosion Costs and Preventive Strategies in the United States. Available online: https://www.nace.org/uploadedfiles/publications/ccsupp.pdf (accessed on 28 December 2018).
4. Ates, M. A review on conducting polymer coatings for corrosion protection. *J. Adhes. Sci. Technol.* **2016**, *30*, 1510–1536. [CrossRef]
5. Burkarter, E.; Saul, C.K.; Thomazi, F.; Cruz, N.C.; Roman, L.S.; Schreiner, W.H. Superhydrophobic electrosprayed PTFE. *Surf. Coat. Technol.* **2007**, *202*, 194–198. [CrossRef]

6. Qian, B.; Shen, Z. Fabrication of superhydrophobic surfaces by dislocation-selective chemical etching on aluminum, copper, and zinc substrates. *Langmuir* **2005**, *21*, 9007–9009. [CrossRef] [PubMed]
7. Meng, L.Y.; Park, S.J. Effect of fluorination of carbon nanotubes on superhydrophobic properties of fluoro-based films. *J. Colloid Interface Sci.* **2010**, *342*, 559–563. [CrossRef]
8. Acatay, K.; Simsek, E.; Ow-Yang, C.; Menceloglu, Y.Z. Tunable, superhydrophobically stable polymeric surfaces by electrospinning. *Angew. Chem.* **2004**, *43*, 5210–5213. [CrossRef]
9. Rivero, P.J.; Garcia, J.A.; Quintana, I.; Rodriguez, R. Design of nanostructured functional coatings by using wet-chemistry methods. *Coatings* **2018**, *8*, 76. [CrossRef]
10. Yan, L.; Li, Y.S.; Xiang, C.B. Preparation of poly(vinylidene fluoride) (PVDF) ultrafiltration membrane modified by nano-sized alumina (Al_2O_3) and its anti-fouling research. *Polymer* **2005**, *46*, 7701–7706. [CrossRef]
11. Sundaram, N.T.K.; Vasudevan, T.; Subramania, A. Synthesis of ZrO_2 nano particles in microwave hydrolysis of Zr (IV) salt solutions-ionic conductivity of PVDF-co-HFP- based polymer electrolyte by the inclusion of ZrO_2 nanoparticles. *J. Phys. Chem. Solids* **2007**, *68*, 264–271. [CrossRef]
12. Hashim, N.A.; Liu, Y.; Li, K. Preparation of PVDF hollow fiber membranes using SiO_2 particles: The effect of acid and alkali treatment on the membrane performances. *Ind. Eng. Chem. Res.* **2011**, *50*, 3035–3040. [CrossRef]
13. Yi, X.S.; Yu, S.L.; Shi, W.X.; Sun, N.; Jin, L.M.; Wang, S.; Zhang, B.; Ma, C.; Sun, L.P. The influence of important factors on ultrafiltration of oil/water emulsion using PVDF membrane modified by nano-sized TiO_2/Al_2O_3. *Desalination* **2011**, *281*, 179–184. [CrossRef]
14. Wu, L.; Tao, C.Y.; Sun, C.X. Preparation and characterization of Fe_3O_4/PVDF magnetic composite membrane. *Acta Phys.-Chim. Sin.* **2004**, *20*, 598–601. [CrossRef]
15. Trigo, C.E.L.; Porto, A.O.; De Lima, G.M. Characterization of CdS nanoparticles in solutions of P(TFE-*co*-PVDF/*co*-Prop)/N,N-dimethylformadine. *Eur. Polym. J.* **2004**, *40*, 2465–2469. [CrossRef]
16. Mohamed, A.M.A.; Jafari, R.; Farzaneh, M. An optimization of superhydrophobic polyvinylidene fluoride/zinc oxide materials using Taguchi method. *Appl. Surf. Sci.* **2014**, *288*, 229–237. [CrossRef]
17. Shelke, V.; Bhole, M.P.; Patil, D.S. Effect of open air annealing on spin coated aluminum doped ZnO nanostructure. *Mater. Chem. Phys.* **2013**, *141*, 81–88. [CrossRef]
18. Shelke, V.; Bhole, M.P.; Patil, D.S. Opto-electrical characterization of transparent conducting sand dune shaped indium doped ZnO nanostructures. *J. Alloys Compd.* **2013**, *560*, 147–150. [CrossRef]
19. Shelke, V.; Bhole, M.P.; Patil, D.S. Open air annealing effect on the electrical and optical properties of tin doped ZnO nanostructure. *Solid State Sci.* **2012**, *14*, 705–710. [CrossRef]
20. Firouzi, A.; Impagnatiello, A.; Del Gaudio, C.; Lamastra, F.R.; Bianco, A.; Montesperelli, G. Electrospun protective self-healing coatings for light alloys: A better understanding of the intrinsic potential of the technology. *J. Appl. Polym. Sci.* **2015**, *132*, 42728. [CrossRef]
21. Tarus, B.; Fadel, N.; Al-Oufy, A.; El-Messiry, M. Effect of polymer concentration on the morphology and mechanical characteristics of electrospun cellulose acetate and poly (vinyl chloride) nanofiber mats. *Alex. Eng. J.* **2016**, *55*, 2975–2984. [CrossRef]
22. El Haleem, S.A.; El Wanees, S.A.; Bahgat, A. Environmental factors affecting the corrosion behaviour of reinforcing steel. VI. Benzotriazole and its derivatives as corrosion inhibitors of steel. *Corros. Sci.* **2014**, *87*, 321–333. [CrossRef]
23. Zampetti, E.; Pantalei, S.; Scalese, S.; Bearzotti, A.; de Cesare, F.; Spinella, C.; Macagnano, A. Biometic sensing layer based on electrospun conductive polymer webs. *Biosens. Bioelectron.* **2011**, *26*, 2460–2465. [CrossRef] [PubMed]
24. Sill, T.J.; von Recum, H.A. Electrospinning: Applications in drug delivery and tissue engineering. *Biomaterials* **2008**, *29*, 1989–2006. [CrossRef]
25. Lee, K.; Kim, H.Y.; La, Y.M.; Lee, D.R.; Sung, N.H. Influence of a mixing solvent with tetrahydrofuran and *N,N*-dimethylformamide on electrospun poly(vinyl chloride) nonwoven mats. *J. Polym. Sci. Part B Polym. Phys.* **2002**, *40*, 2259–2268. [CrossRef]
26. Megelski, S.; Stephens, J.S.; Chase, D.B.; Rabolt, J.F. Micro and nanostructured surface morphology on electrospun polymer fibers. *Macromolecules* **2002**, *35*, 8456–8466. [CrossRef]
27. Ma, M.; Mao, Y.; Gupta, M.; Gleason, K.; Rutledge, G.C. Superhydrophobic fabrics produced by electrospinning and chemical vapor deposition. *Macromolecules* **2005**, *38*, 9742–9748. [CrossRef]

28. Wenzel, R.N. Surface roughness and contact angle. *J. Phys. Chem.* **1949**, *53*, 1466–1467. [CrossRef]
29. Cassie, A.B.D.; Baxter, S. Wettability of porous surfaces. *Trans. Faraday Soc.* **1944**, *40*, 546–551. [CrossRef]
30. Gao, L.; McCarthy, T.J. An Attempt to correct the faulty intuition perpetuated by the Wenzel and Cassie "laws". *Langmuir* **2009**, *25*, 7249–7255. [CrossRef]
31. Erbil, H.Y. The debate on the dependence of apparent contact angles on drop contact area or three-phase contact line: A review. *Surf. Sci. Rep.* **2014**, *69*, 325–365. [CrossRef]
32. Yuan, Y.; Choi, S.O.; Kim, J. Analysis of contact area between water and irregular fibrous surface for prediction of wettability. *RSC Adv.* **2016**, *6*, 73313–73322. [CrossRef]
33. Giacomello, A.; Meloni, S.; Chinappi, M.; Casciola, C.M. Cassie–Baxter and Wenzel states on a nanostructured surface: Phase diagram, metastabilities, and transition mechanism by atomistic free energy calculations. *Langmuir* **2012**, *28*, 10764–10772. [CrossRef]
34. Milne, A.J.B.; Amirfazli, A. The Cassie equation: How it is meant to be used. *Adv. Colloid Interface Sci.* **2012**, *170*, 48–55. [CrossRef]
35. Szewczyk, P.K.; Ura, D.P.; Metwally, S.; Knapczyk-Korczak, J.; Gajek, M.; Marzec, M.M.; Bernasik, A.; Stachewicz, U. Roughness and fiber fraction dominated wetting of electrospun fiber-based porous meshes. *Polymers* **2019**, *11*, 34. [CrossRef]
36. Homaeigohar, S.; Koll, J.; Lilleodden, E.T.; Elbahri, M. The solvent induced interfiber adhesion and its influence on the mechanical and filtration properties of polyethersulfone electrospun nanofibrous microfiltration membranes. *Sep. Purif. Technol.* **2012**, *98*, 456–463. [CrossRef]
37. Zhang, L.; Liu, L.; Pan, F.; Wang, D.; Pan, Z. Effects of heat treatment on the morphology and performance of PSU electrospun nanofibrous membrane. *J. Eng. Fibers Fabr.* **2012**, *7*, 12–13. [CrossRef]
38. Jadhav, N.R.; Gaikwad, V.L.; Nair, K.J.; Kadam, H.W. Glass transition temperature: Basics and application in pharmaceutical sector. *Asian J. Pharm.* **2009**, *3*, 82–89. [CrossRef]
39. Asmatulu, R.; Ceylan, M.; Nuraje, N. Study of superhydrophobic electrospun nanocomposite fibers for energy systems. *Langmuir* **2011**, *27*, 504–507. [CrossRef] [PubMed]
40. Scully, J.; Budiansky, N.; Tiwary, Y.; Mikhailov, A.; Hudson, J. An alternate explanation for the abrupt current increase at the pitting potential. *Corros. Sci.* **2008**, *50*, 316–324. [CrossRef]

coatings

Article

Effect of Incorporating MoS_2 in Organic Coatings on the Corrosion Resistance of 316L Stainless Steel in a 3.5% NaCl Solution

Min-Sung Hong [1,†], Yunjeong Park [2,†], Jung Gu Kim [1,*] and Kyunghoon Kim [2,*]

1 School of Advanced Material Science and Engineering, Sungkyunkwan University (SKKU), Suwon 16419, Korea; smith803@skku.edu
2 School of Mechanical Engineering, Sungkyunkwan University (SKKU), Suwon 16419, Korea; djy828@skku.edu
* Correspondence: kimjg@skku.edu (J.G.K.); kenkim@skku.edu (K.K.); Tel.: +82-31-290-7360 (J.G.K.); +82-31-290-7432 (K.K.)
† These authors contributed equally to this study.

Received: 11 December 2018; Accepted: 11 January 2019; Published: 15 January 2019

Abstract: This study discusses a new coating method to protect 316L stainless steel (SS) from pitting corrosion in high chloride environments. The SS surface was coated using a simple, eco-friendly method, and sunflower oil (SunFO) was used as a base coating and binder for molybdenum disulfide (MoS_2). The coated surface was observed using scanning electron microscopy (SEM) with an energy dispersive spectrometer (EDS) and X-ray diffraction (XRD). Corrosion behavior was examined by open-circuit potential (OCP) measurement and electrochemical impedance spectroscopy (EIS) in an 3.5% NaCl solution. The SunFO coating with MoS_2 showed the highest corrosion resistance and coating durability during the immersion time relative to the SunFO coating and bare 316L SS. The increased corrosion resistance is thought to be because of the interactions with the aggregations of the SunFO lamellar structure and MoS_2 in the coating film, which acted as a high order layer barrier providing protection from the metals to electrolytes.

Keywords: stainless steel; sunflower oil; molybdenum disulfide (MoS_2); organic coating; corrosion resistance; electrochemical impedance spectroscopy (EIS)

1. Introduction

Austenitic stainless steel (SS) has excellent corrosion resistance in various environments, so they are used as materials in multiple applications such as pipelines, pumps, and structural steels in many industries [1]. Among them, 316L SS has good mechanical and welding properties with high corrosion resistance; therefore, it is regarded as one of the most effective materials in the field of corrosion [2,3]. The high corrosion resistance of 316L SS comes from a thin chromium oxide film which has high stability against corrosion [4]. In marine atmospheric environments, however, it often suffers from pitting-type corrosion due to exposure to stagnant seawater and through the deposition of airborne sea salts, which have high humidity and high chloride concentration [5]. When pitting corrosion is initiated, the pit propagates aggressively because the pH within the pit inside turns acidic due to the generation of $CrCl_3$, which is called the autocatalytic mechanism [6]. Moreover, the pitting corrosion of SS could cause stress corrosion cracking (SCC) which results in unpredictable, brittle fracture [7]. For these reasons, corrosion protection of stainless steel in high chloride environments is a major concern in many industries. As such, several corrosion mitigation methods are being studied and developed in various fields.

Alkyd coatings are extensively used for the surface coatings as binders and adhesives. This class of coating is generated from polyols, dibasic and fatty acids or oils by condensation polymerization [8].

Alkyds have attracted significant attention among coating materials because they are lower cost and incur fewer film defects during applications. Recently, vegetable oils have been highlighted as a new effective organic coating that are nontoxic, nondepletable, domestically abundant, nonvolatile and a biodegradable resource [9]. The vegetable oils are triacylglycerols of fatty acids with high degrees of unsaturated sites which can have the ability to polymerize via cross-linking under certain conditions [10]. Polar molecules present in the oils can be absorbed on metal surfaces and form the corresponding metal oxides, which will enhance the stability of passivation and promote adhesion [9]. Sunflower oil (SunFO) coating is reported to be an effective inhibitor of corrosion for carbon steel, which is likely due to the lamellar-like layered structures of the organic film [9,11]. However, there are few studies surrounding the corrosion inhibition mechanism of the SunFO coating for SS and methods for synthesizing the SunFO coating using other effective materials. Therefore, in this study, the sunflower oil is selected as a base coating and binder for 316L SS.

The two-dimensional (2D) materials have been extensively researched within the context of several applications as coating materials because of their interesting, atomically thin, physical, chemical and electrical properties [12,13]. Among 2D materials, molybdenum disulfide (MoS_2) has been widely applied as a lubrication and thin film protection coating materials because of its tribological and corrosion resistance properties [14,15]. Moreover, MoS_2 remains stable in various solvents and oxygenated environments, and it can also withstand high temperatures and pressures [16,17]. Currently, however, many surface modifications, such as chemical vapor deposition (CVD), water transfer are being phased out due to their high cost, lengthy processing time, low output and harmful effects on environments. This has motivated current research, which is being conducted to improve the stability and corrosion resistance with a mixture consisting of MoS_2 particles and several organic coating materials without imparting the toxic effects [18].

This study discusses a new coating method to protect 316L SS from the pitting corrosion when exposed to a 3.5% NaCl solution. Scanning electron microscopy (SEM) with energy dispersive spectrometry (EDS) was used to observe the dispersion of MoS_2 on the SS surface and cross section. In addition, X-ray diffraction (XRD) measurements were performed to evaluate the coating materials on SS. After that, the electrochemical properties of organic coatings according to MoS_2 were evaluated by using open-circuit potential (OCP) measurement and electrochemical impedance spectroscopy (EIS) tests.

2. Materials and Methods

2.1. Specimens and Solution Preparation

The chemical composition of the 316L SS is given in Table 1. For the electrochemical tests and the coating procedure, the surface of the specimen was polished with 2000-grit silicon carbide (SiC) paper, rinsed with ethanol and then dried with N_2 gas.

Table 1. Chemical Composition of 316L stainless steel (wt.%).

Elements	Composition
Fe	Balance
C	0.03 Max.
Cr	16–18
Ni	10–14
Mo	2–3
S	0.03 Max.

Figure 1 shows a schematic diagram of the experimental procedures and expected coating structure. A 1 mg amount of MoS_2 particles (commercially purchased, Sigma-Aldrich, St. Louis, MI, USA) was mixed with 10 mL of sunflower oil (commercially purchased) in 20 mL beakers. The mixture was sonicated and then stirred at 1000 rpm at room temperature for 3 h. Then, 100 μL of the mixture was dropped on the SS substrate and heated at 275 °C for 10 min on a hot plate and slowly cooled. At this temperature, the triglycerides which is a main composition of sunflower oil undergo polymerization through oxidation [10]. The SunFO with MoS_2 coatings were presumed to have a layered oil structure which contained evenly distributed MoS_2 particles.

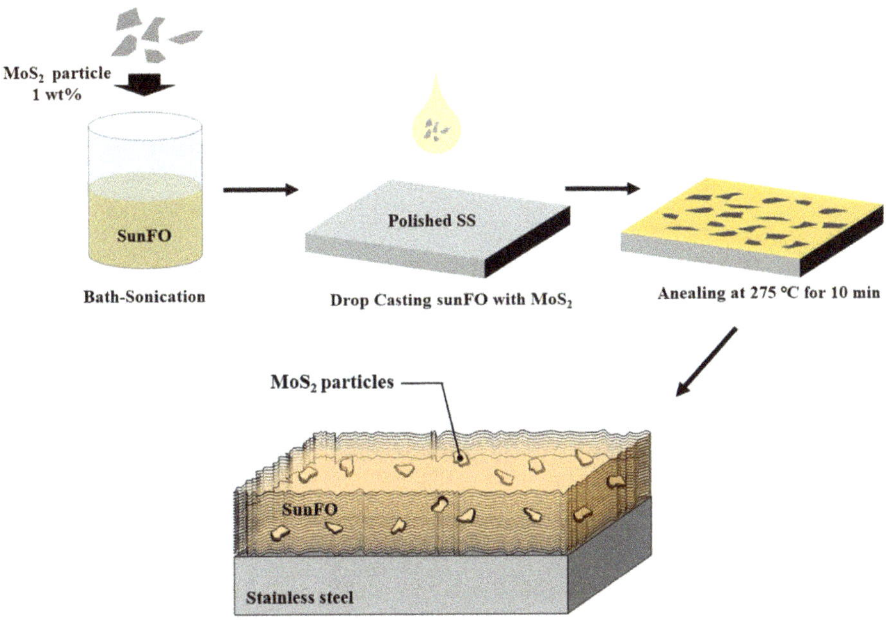

Figure 1. Schematic diagram of the experimental procedures and expected coating structure.

To observe the cross section of the coated specimens, the cross section was polished with 2000-grit silicon carbide (SiC) paper. All electrochemical experiments were conducted in a 3.5% NaCl solution at ambient temperature.

2.2. *Surface Analysis*

The surface morphology and the cross-sectional image of the specimens were observed using SEM/EDS (JSM-7600F, Jeol Ltd., Tokyo, Japan) to verify the dispersion of MoS_2 in organic coating. X-ray diffraction (XRD, D8 Advance, Bruker Co., Karlsruhe, Germany) measurements were performed on the specimens to identify the effective bonding within the organic coating, MoS_2 and 316L SS. The XRD analysis of the coated specimens was conducted to confirm the crystalline properties of the coating film using Cu Kα radiation (λ = 1.54178 Å) in the 2θ range 0–60° at a scan rate of 0.017° 2θ.

2.3. *Electrochemical Investigation Method*

All electrochemical experiments were performed using a three-electrode system in a 1000 mL Pyrex glass corrosion cells connected to an electrochemical apparatus (VSP 300, Bio-Logic SAS, Seyssinet-Pariset, France). The test specimens were connected to a working electrode, graphite rods were used as the counter electrode and a saturated calomel electrode (SCE) was used as the reference electrode. The exposed area of the test specimen to the electrolyte was 0.25 cm^2 (0.5 cm × 0.5 cm). An initial open-circuit potential (OCP) was established within 3 h to carry out the entire electrochemical test. The electrochemical impedance spectroscopy (EIS, VSP 300, Bio-Logic SAS, Seyssinet-Pariset, France) was carried out with an amplitude of 20 mV in a frequency range from 100 kHz to 10 mHz. The EIS tests were performed at 3, 12, 36 and 63 h to investigate variations of the coating durability in corrosive media. The impedance plots were interpreted on the basis of equivalent circuits using a suitable fitting procedure through the ZSimpWin software (ZsimpWin 3.20).

3. Results

3.1. Surface Analysis

The SEM/EDS analysis is used to probe the morphologies of the SunFO coated SS. The addition of the pristine MoS_2 in the SunFO significantly impacted the surface morphology. As shown in Figure 2, the SunFO with the MoS_2 coating film shows the presence of a large number of MoS_2 flakes which averaged an even layer 1–3 μm thick on the entire surface (Figure 2a–d). The EDS data indicated that the surface consisted of Mo, S, Cr, Ni, and Fe corresponding to particles of MoS_2 and the surface of the SS (Figure 2e).

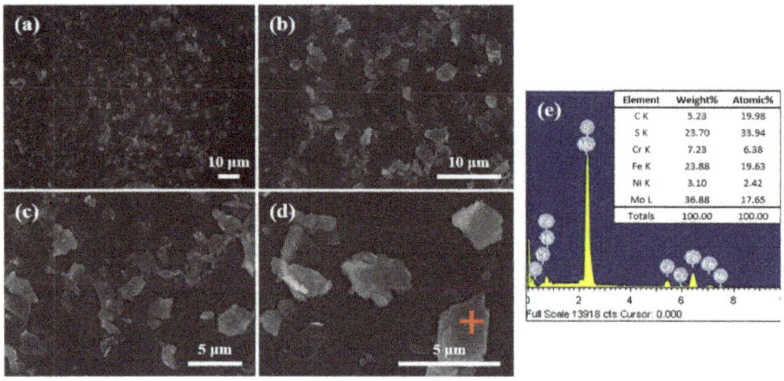

Figure 2. Surface morphology of sunflower oil (SunFO) with molybdenum disulfide (MoS_2) coating on the stainless steel (SS) observed by SEM and EDS: (**a**) 1000×; (**b**) 3000×; (**c**) 5000×; (**d**) 10000×; (**e**) EDS results on the red spot.

Figure 3 shows cross-sectional images of SunFO with the MoS_2 coating specimen. The thickness of the coating was verified at about 3–4 μm and shows uniform and dense surface features. According to the EDS mapping results (Figure 3b,c), it was revealed that the MoS_2 particle existed in the SunFO coating layer. However, it was difficult to observe the layered structure of the oil, so the SEM analysis was re-conducted on an unpolished condition. In Figure 4, a layered structure was observed, and it was demonstrated that the coating structure of the specimen corresponded to the expectation outlined in Figure 1.

The XRD patterns in Figure 5 show the crystalline nature of the MoS_2 and substrate. In previous reports, the diffraction pattern of bulk MoS_2 has strong and sharp peaks which correspond to (100), (002), (100), (103), (105), and (110) planes [19]. After SunFO with MoS_2 coating, the MoS_2 peaks were only detected via the (002) peak corresponding to the angle of 15° because the MoS_2 was oriented to a layered structure on the SS. The additional diffraction peaks at 44.5° and 19.6° indicate crystalline iron in the SS substrate and SunFO.

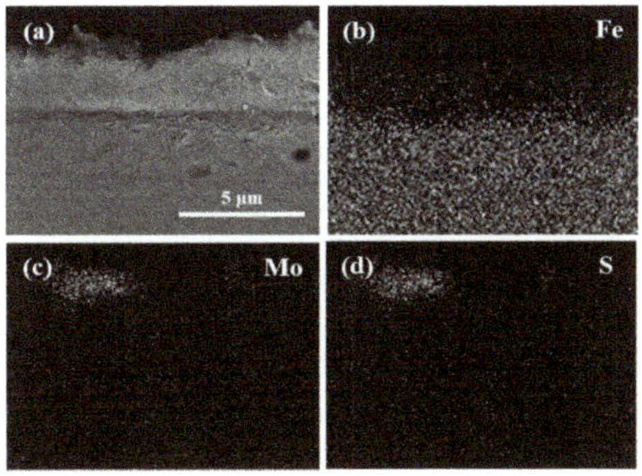

Figure 3. Cross sectional images of SunFO with MoS_2 coating on the SS observed by SEM and EDS: (**a**) 10000× (Cross-section of the specimen); (**b**) Fe observed by EDS mapping; (**c**) Mo observed by EDS mapping; (**d**) S observed by EDS mapping.

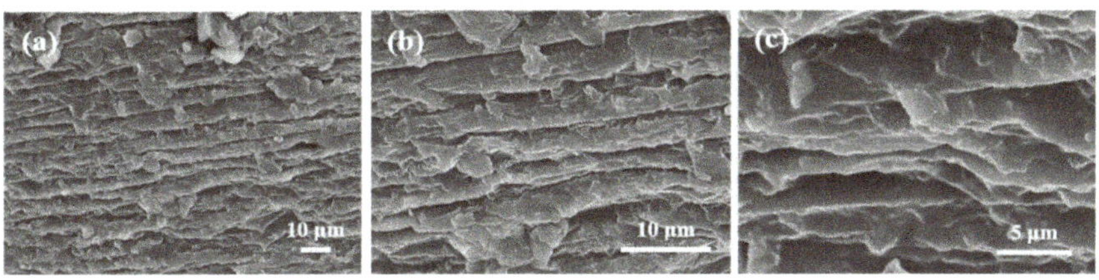

Figure 4. SEM images of unpolished SunFO with MoS_2 coating on the SS observed by SEM: (**a**) 1000×; (**b**) 3000×; (**c**) 5000×.

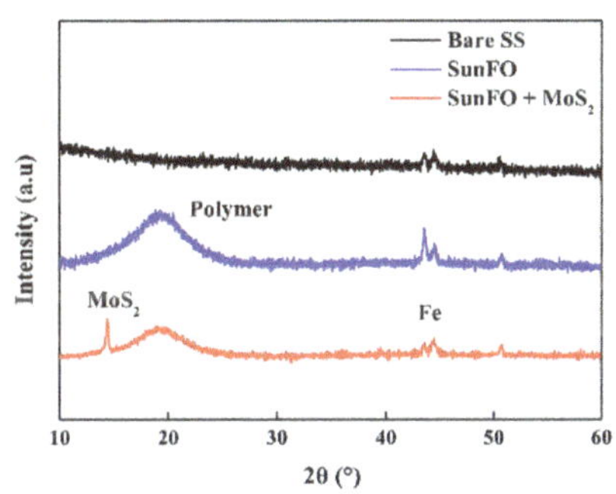

Figure 5. XRD results of (**a**) Bare SS; (**b**) SunFO coated SS; (**c**) SunFO with MoS_2 coated SS.

3.2. Electrochemical Analysis (EIS Tests)

The impedance spectra were obtained in the form of Nyquist plots of the data from the three-electrode system as shown in Figure 6. The tests were conducted with the following specimens, bare, SunFO coating and SunFO with a MoS_2 coating at room temperature. The specimens were tested in a 3.5% NaCl solution for a total of 63 h. The Nyquist plot consisted of a depressed capacitive loop at high frequency values [20,21]. As shown in Figure 6a, the capacitive loop of bare SS increased as time progresses, because the passive film on the surface goes to stable in the solution [22–24]. While, in case of the SunFO coated SS, the loops decreased with immersion time, which was caused by a deterioration of oil coating (Figure 6b) [25,26]. The capacitive loop of SunFO coating with MoS_2 on SS had little change during the test period, so it was regarded as a non-time-dependent effect in the solution as shown in Figure 6c.

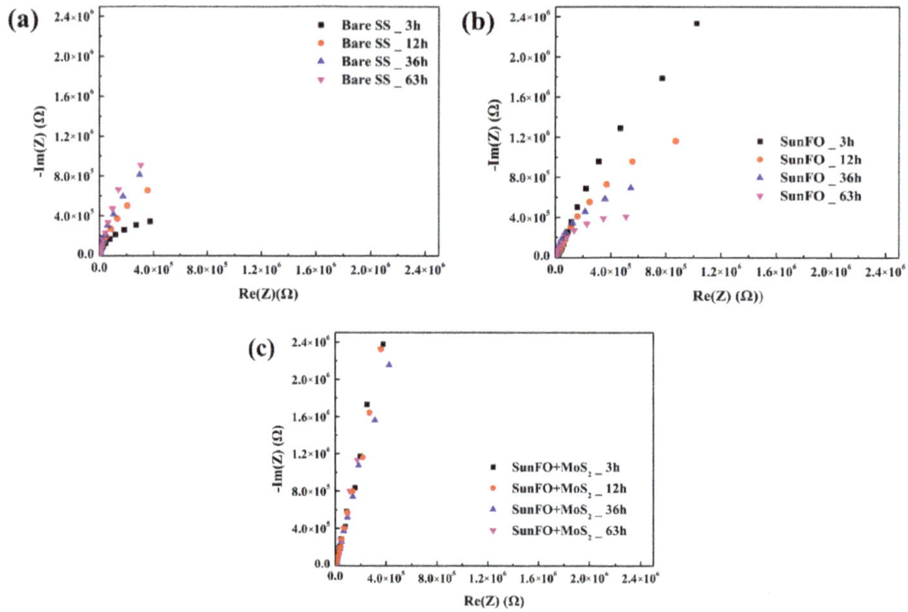

Figure 6. Electrochemical impedance spectroscopy (EIS) results in the form of Nyquist plot in 3.5% NaCl solution; (**a**) Bare SS; (**b**) SunFO coated SS; (**c**) SunFO + MoS_2 coated SS.

In Figure 7a, the capacitive loop of bare SS was the smallest initially, however it displayed a reversed trend compared with the plot of the SunFO coated SS at the final time (Figure 7b). The SunFO coating with MoS_2 on the surface maintained the loop tendency during the entire immersion time. It can be also verified in the OCP graph seen in Figure 8. The SunFO with MoS_2 coating SS revealed very stable potential, which obviously differed from the bare and SunFO coated SS. Moreover, the potential of SunFO coated SS continuously decreased and reversed compared with the potential of bare specimen after about 28 h, which was the same tendency observed in the time-dependent variation of the Nyquist plot.

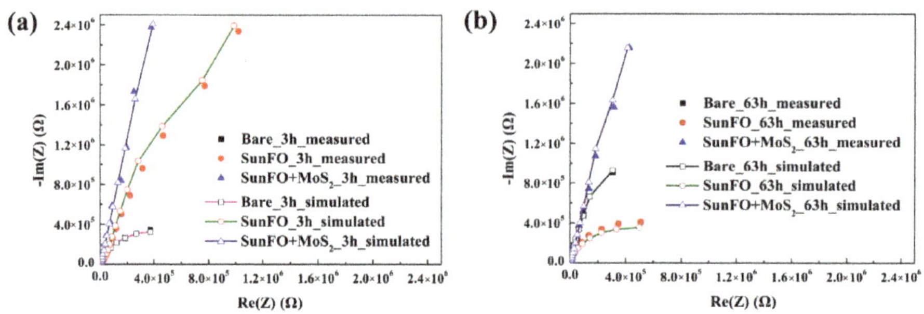

Figure 7. Time-dependent EIS results in the form of Nyquist plot in 3.5% NaCl solution: (**a**) initial stage (3 h); (**b**) final stage (63 h).

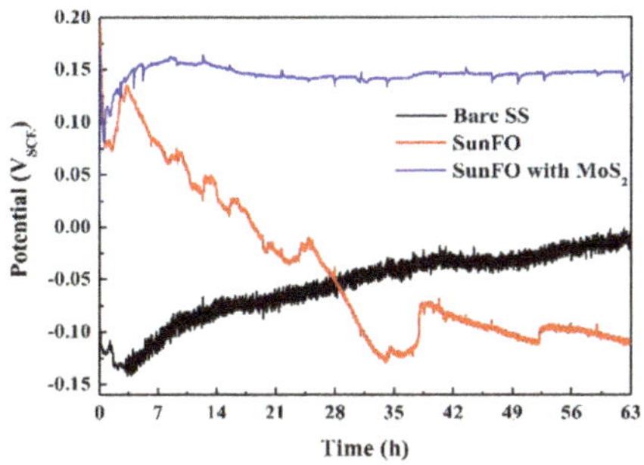

Figure 8. Open-circuit potential of specimens during the immersion time.

Figure 9 presents the EIS results in the form of Bode phase plots during immersion for 63 h. The Bode phase plot provided more clear description of the electrochemical frequency-dependent behavior than did the Nyquist plot, where the frequency values are implicit [27]. The high frequency spectra detects local surface defects, whereas the low frequency spectra detects the processes within the film and at the metal/film interface, respectively [28,29]. As shown in Figure 9a, in case of the bare SS, the phase angle maximum is slightly increased, and the width of the graph is wider in the low frequency region according to the test time. It means the passive film on the surface goes to more uniform and thicker [30–33]. In the case of SunFO coated SS, there was evidence of variations in surface conditions (Figure 9b). The initial stage of the specimen had two-time constants, but it changed to one-time constant at the final stage [34]. In addition, the Bode plot of SunFO coated SS has a low phase angle maximum and narrow shoulder width. This indicated that the SunFO coating had a poor protective property and has many defects [35–38]. The specimen coated with SunFO and MoS_2 did not exhibit strong time-dependent behavior in the graphs, however it had a very wide frequency area as shown in Figure 9c. This is thought to be due the maintenance of a thick and uniform film throughout the entire test period.

Figure 10a shows the equivalent electrical circuits for bare SS, which had a passive film, and was used to analyze the results of the EIS tests (one-time constant circuit) [27]. In this figure, R_s is the solution resistance, *CPE*1 is the dielectric strength of the film and water absorbed by the film, and R_{film} is the electrical resistance resulting from the formation of an ionic conduction path through the pores in the film. The capacitance generated by the metal dissolution reaction and by the electric double layer at the electrolyte/substrate interface is designated by *CPE*2, and R_{ct} is the resistance caused by the metal dissolution reaction. In the case of SunFO coated specimens, however, more circuit parameters should be added because the oil coating generates a new layer which forms a two-layered film (two-time constant circuit) [39]. As shown in Figure 10b, $R_{coating}$ is the SunFO coating resistance and *CPE*3 is the dielectric strength of the SunFO coating. The ZSimpWin program of the defined equivalent circuits was used to fit the EIS data to determine the optimized values for the resistance parameters, which are presented in Table 2. The R_{film} and R_{ct} of bare SS increased according to the immersion time. These values indicated an increase of the passive film's stability, which was the same tendency identified in the Nyquist and Bode plot. The film and coating capacitance (C_{film}, $C_{coating}$) is described by the expression [20,40]:

$$C = \frac{\varepsilon A\prime}{d} \tag{1}$$

where $A\prime$ is the surface area of specimen, ε is the dielectric constant, and d is thickness of the passive film in solution. This equation suggests that a decrease of C_{dl} is related to an increase of passive film's thickness. The C_{film} of bare SS decreased with time, therefore this indicated that the passive film on the surface becomes thicker at final stage. In the case of the SunFO coated specimen, there were too many error values when it was applied to a two-time constant circuit after the initial stage. As shown in the Bode plot, two-time constants were displayed only at the initial stage and one-time constant was shown after that. In other words, the SunFO coating on the SS surface became degraded prior to 12 h. For this reason, the simulation was conducted to the point of one-time constant after the initial stage. According to the data shown in Table 2, R_{film} and R_{ct} decreased according to immersion time, which meant the film had deteriorated. Especially, the R_{film} and C_{film} rapidly decreased after 36 h which indicates that a thick and porous non-protective layer was generated on the surface [41–43]. The SunFO with MoS_2 coated SS showed an obvious two-time constant, which fit well with the two-time circuit. Both values, $R_{coating}$ and R_{ct}, increased according to the test period, and the $R_{coating}$ value at 63 h was markedly greater than the R_{film} of bare SS at the same time, which indicated excellent protective film properties. Moreover, the R_{film} value remained very small relative to that of bare SS. It is presumed that the SunFO coating with the MoS_2 blocks the electrolyte and oxygen so that the passive film could not form.

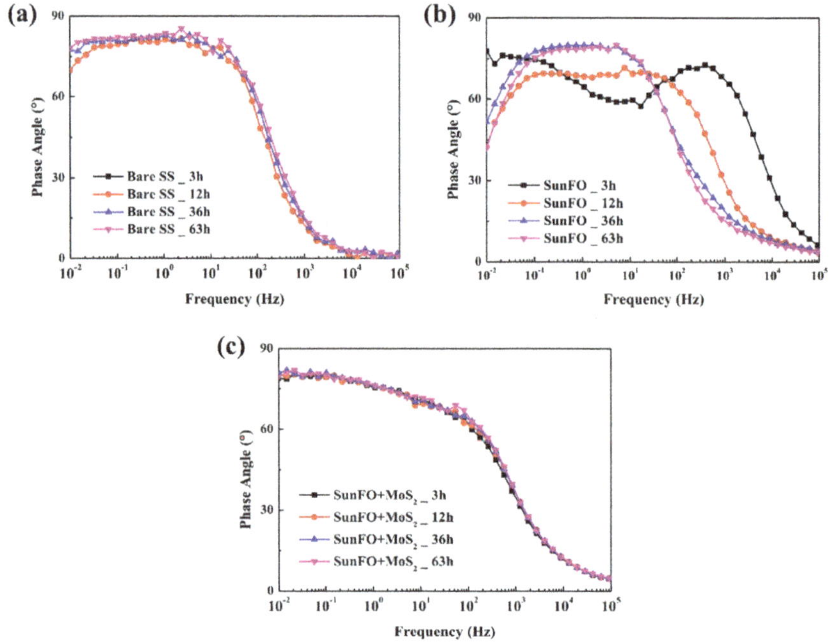

Figure 9. EIS results in the form of Bode plot in 3.5% NaCl solution: (**a**) Bare SS; (**b**) SunFO coated SS; (**c**) SunFO + MoS_2 coated SS.

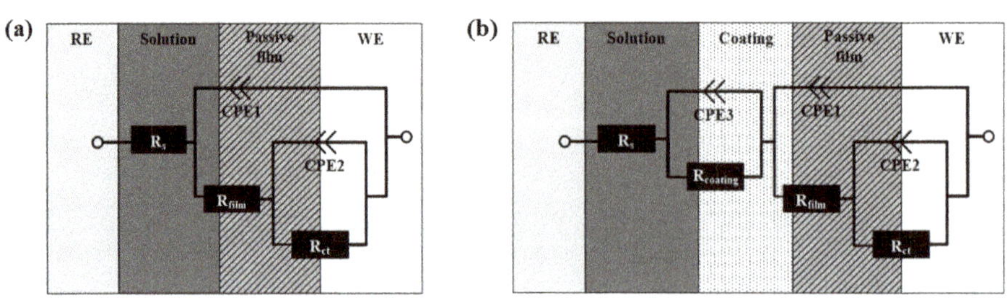

Figure 10. An equivalent circuit used to fit the results of the EIS tests: (**a**) one-time constant circuit; (**b**) two-time constant circuit.

Table 2. EIS results of specimens according to the coating materials.

Specimen	Time	R_s (Ω cm^2)	CPE1 (C_{film}) (F/cm^2)	n_1	R_{film} (Ω cm^2)	CPE2 (C_{dl}) (F/cm^2)	n_2	R_{ct} (Ω cm^2)	n_3	CPE3 ($C_{coating}$) (F/cm^2)	$R_{coating}$ (Ω cm^2)
Bare	3	118.9	1.37×10^{-5}	0.92	9.85×10^{4}	2.06×10^{-6}	0.37	1.01×10^{6}	–	–	–
	12	118.2	1.37×10^{-5}	0.91	5.65×10^{5}	1.88×10^{-6}	0.61	1.83×10^{6}	–	–	–
	36	119.5	1.34×10^{-7}	0.91	3.91×10^{6}	3.74×10^{-6}	0.80	2.14×10^{6}	–	–	–
	63	120.1	4.73×10^{-6}	0.91	2.12×10^{7}	7.61×10^{-6}	0.83	2.12×10^{7}	–	–	–
SunFO	3	120.4	7.72×10^{-7}	0.93	1.22×10^{4}	1.88×10^{-6}	0.83	6.33×10^{7}	0.80	2.31×10^{-4}	4.38×10^{-6}
	12	127.6	2.39×10^{-6}	0.88	3.19×10^{4}	3.61×10^{-6}	0.77	4.88×10^{5}	–	–	–
	36	134.1	4.55×10^{-7}	0.86	181.5	7.39×10^{-6}	0.91	1.58×10^{6}	–	–	–
	63	126.9	7.66×10^{-8}	0.81	427.7	4.80×10^{-6}	0.94	9.05×10^{5}	–	–	–
SunFO + MoS_2	3	135.7	4.80×10^{-5}	0.56	223.5	6.20×10^{-7}	0.97	1.01×10^{4}	0.90	5.12×10^{-6}	2.836×10^{9}
	12	135.1	5.05×10^{-5}	0.56	238.5	4.65×10^{-7}	1	1.48×10^{4}	0.90	5.29×10^{-6}	3.05×10^{10}
	36	141.4	3.76×10^{-5}	0.63	6742	4.50×10^{-5}	0.77	3.48×10^{7}	0.89	4.96×10^{-6}	1.34×10^{14}
	63	140.1	4.52×10^{-5}	0.60	6855	2.69×10^{-5}	0.98	1.24×10^{9}	0.89	5.07×10^{-6}	1.19×10^{16}

The SunFO is a complex mixture of triacylglycerol consisting of tri-esters of glycerol and fatty acid [44,45]. At high temperature, the unsaturated fatty acid of SunFO undergoes oxidation reactions and cross links with the SS. According to the XRD results (Figure 5), the SunFO layer has crystalline peaks at 19.6° which means the liquid states of the fat were crystallized on the SS. According to the expected coating structure (Figure 1), the SunFO coating, which has fatty acid hydrocarbon chains, makes lamellar structures as identified by the subcell concept [9,11]. However, the lamellar structures of the SunFO are aggregated structures that have a number of defects. In the SunFO with MoS_2 coating film, the MoS_2 has layered structures that, during the coating process on SS, are well layered [8,9,11,46]. The interactions with aggregations of the SunFO lamellar structure and layered MoS_2 in coating the films acts as a high ordered layer barrier for the protection of the metals from electrolytes. The defects of both materials act as active centers for molecular adsorption and functionalization, therefore the combination of both layered materials could reduce the defects and build denser film [47]. Additionally, previous studies have reported that MoS_2 particles impart a negative effect on the corrosion because of their cathodic partial reactions during the corrosion process which leads to the destruction of the surface film [48]. Moreover, the electrochemical potential of MoS_2 particle is higher than the 316L stainless steel, so it could generate galvanic cell between steel and MoS_2 which accelerates corrosion reaction on the steel surface [49]. However, in case of the MoS_2 with the SunFO film, the lamellar structure of the oil acted as an electrical insulating barrier between MoS_2 and steel surface, which prevents the galvanic corrosion and improves the corrosion resistance.

4. Conclusions

This study evaluated the reinforcement effect of MoS_2 in organic coatings on 316L SS in 3.5% NaCl solution. Firstly, SEM with EDS and XRD methods were performed to verify the MoS_2 distribution on the SS surface. Through the experiments, the following results were drawn:

- The SEM images of SunFO with MoS_2 coating film showed the presence of a large number of MoS_2 flakes which averaged 1–3 μm evenly on the entire surface. The XRD results showed that the crystalline nature and orientation of the MoS_2 have strong and sharp peaks, which are the (002) peak corresponding to the angle of 15°.
- In the EIS results, the bare SS showed stable passive film generation during the test time. The SunFO with the MoS_2 coating on the SS surface showed the largest coating resistance and durability throughout the immersion time compared to SunFO coating and bare specimen. This is because the interaction with the aggregations of the SunFO lamellar structure and MoS_2 in coating film acted as a high order layer barrier to protect the metals against electrolytes.

Consequently, the incorporating MoS_2 in organic coatings could considerably improve the localized corrosion resistance of stainless steel in high chloride environment.

Author Contributions: Experiments Conception and Design, M.-S.H., Y.P., J.G.K. and K.K.; Experiment Performance, M.-S.H. and Y.P.; Data Analysis, M.-S.H., Y.P., J.G.K. and K.K.; Manuscript Writing, M.-S.H., Y.P., J.G.K. and K.K.

Funding: This research was supported by the Basic Science Research Program through the National Research Foundation of Korea (NRF) funded by the Ministry of Science, ICT & Future Planning (NO. NRF-2017R1C1B2011750). This research was supported by the program for fostering next-generation researchers in engineering of National Research Foundation of Korea (NRF) funded by the Ministry of Science and ICT (No. 2017H1D8A2031628). This work was supported by NRF (National Research Foundation of Korea) Grant funded by the Korean Government (NRF-2018-Fostering Core Leaders of the Future Basic Science Program/Global Ph.D. Fellowship Program) (No. NRF-2018H1A2A1062418).

Conflicts of Interest: The authors declare no conflict of interest.

References

1. Uhlig, H.H.; King, C. Corrosion and corrosion control. *J. Electrochem. Soc.* **1972**, *119*, 327C. [CrossRef]
2. Galliano, F.; Landolt, D. Evaluation of corrosion protection properties of additives for waterborne epoxy coatings on steel. *Prog. Org. Coat.* **2002**, *44*, 217–225. [CrossRef]
3. Kain, R.M. *Metal Handbook*; ASM International: Novelty, OH, USA, 1987; Volume 13.
4. Fossati, A.; Borgioli, F.; Galvanetto, E.; Bacci, T. Corrosion resistance properties of glow-discharge nitrided AISI 316L austenitic stainless steel in NaCl solutions. *Corros. Sci.* **2006**, *48*, 1513–1527. [CrossRef]
5. Streicher, M. Pitting corrosion of 18Cr-8Ni stainless steel. *J. Electrochem. Soc.* **1956**, *103*, 375–390. [CrossRef]
6. Uhlig, H.H. *Uhlig's Corrosion Handbook*; John Wiley & Sons: New York, NY, USA, 2011.
7. Tsai, W.-T.; Chen, M.-S. Stress corrosion cracking behavior of 2205 duplex stainless steel in concentrated NaCl solution. *Corros. Sci.* **2000**, *42*, 545–559. [CrossRef]
8. Erciyes, A.T.; Erkal, F.S.; Kalipci, A. Oil-modified alkyd type resin based on secondary esters of castor oil. *J. Coat. Technol.* **1993**, *65*, 73–78.
9. Lligadas, G.; Ronda, J.C.; Galia, M.; Cadiz, V. Renewable polymeric materials from vegetable oils: A perspective. *Mater. Today* **2013**, *16*, 337–343. [CrossRef]
10. Alam, M.; Akram, D.; Sharmin, E.; Zafar, F.; Ahmad, S. Vegetable Oil Based Eco-Friendly Coating Materials: A Review Article. *Arabian J. Chem.* **2014**, *7*, 469–479. [CrossRef]
11. Singhbabu, Y.N.; Sivakumar, B.; Singh, J.K.; Bapari, H.; Pramanick, A.K.; Sahu, R.K. Efficient anti-corrosive coating of cold-rolled steel in a seawater environment using an oil-based graphene oxide ink. *Nanoscale* **2015**, *7*, 8035–8047. [CrossRef]
12. Balakrishnan, T.; Sathiyanarayanan, S.; Mayavan, S. Advanced anticorrosion coating materials derived from sunflower oil with bifunctional properties. *ACS Appl. Mater. Interfaces* **2015**, *7*, 19781–19788. [CrossRef]
13. Bhimanapati, G.R.; Lin, Z.; Meunier, V.; Jung, Y.; Cha, J.; Das, S.; Xiao, D.; Son, Y.; Strano, M.S.; Cooper, V.R. Recent advances in two-dimensional materials beyond graphene. *ACS Nano* **2015**, *9*, 11509–11539. [CrossRef] [PubMed]
14. Butler, S.Z.; Hollen, S.M.; Cao, L.; Cui, Y.; Gupta, J.A.; Gutiérrez, H.R.; Heinz, T.F.; Hong, S.S.; Huang, J.; Ismach, A.F. Progress, challenges, and opportunities in two-dimensional materials beyond graphene. *ACS Nano* **2013**, *7*, 2898–2926. [CrossRef] [PubMed]
15. Spear, J.C.; Ewers, B.W.; Batteas, J.D. 2D-nanomaterials for controlling friction and wear at interfaces. *Nano Today* **2015**, *10*, 301–314. [CrossRef]
16. Britto, R.J.; Benck, J.D.; Young, J.L.; Hahn, C.; Deutsch, T.G.; Jaramillo, T.F. Molybdenum disulfide as a protection layer and catalyst for gallium indium phosphide solar water splitting photocathodes. *J. Phys. Chem. Lett.* **2016**, *7*, 2044–2049. [CrossRef]
17. Choi, J.; Zhang, H.; Du, H.; Choi, J.H. Understanding solvent effects on the properties of two-dimensional transition metal dichalcogenides. *ACS Appl. Mater. Interfaces* **2016**, *8*, 8864–8869. [CrossRef]
18. Bandaru, N.; Kumar, R.S.; Sneed, D.; Tschauner, O.; Baker, J.; Antonio, D.; Luo, S.-N.; Hartmann, T.; Zhao, Y.; Venkat, R. Effect of pressure and temperature on structural stability of MoS_2. *J. Phys. Chem. C* **2014**, *118*, 3230–3235. [CrossRef]
19. Shankara, A.; Menezes, P.L.; Simha, K.; Kailas, S.V. Study of solid lubrication with MoS_2 coating in the presence of additives using reciprocating ball-on-flat scratch tester. *Sadhana* **2008**, *33*, 207–220. [CrossRef]

20. Joensen, P.; Crozier, E.; Alberding, N.; Frindt, R. A study of single-layer and restacked MoS_2 by X-ray diffraction and X-ray absorption spectroscopy. *J. Phys. C Solid State Phys.* **1987**, *20*, 4043. [CrossRef]
21. Jones, D.A. *Principles and Prevention of Corrosion*, 2nd ed.; Prentice Hall: Upper Saddle River, NJ, USA, 1996.
22. Hong, T.; Sun, Y.H.; Jepson, W.P. Study on corrosion inhibitor in large pipelines under multiphase flow using EIS. *Corros. Sci.* **2002**, *44*, 101–112. [CrossRef]
23. Boissy, C.; Alemany-Dumont, C.; Normand, B. EIS evaluation of steady-state characteristic of 316L stainless steel passive film grown in acidic solution. *Electrochem. Commun.* **2013**, *26*, 10–12. [CrossRef]
24. Mohammadi, F.; Nickchi, T.; Attar, M.; Alfantazi, A. EIS study of potentiostatically formed passive film on 304 stainless steel. *Electrochim. Acta* **2011**, *56*, 8727–8733. [CrossRef]
25. Simoes, A.; Ferreira, M.; Rondot, B.; Da Cunha Belo, M. Study of passive films formed on AISI 304 stainless steel by impedance measurements and photoelectrochemistry. *J. Electrochem. Soc.* **1990**, *137*, 82–87. [CrossRef]
26. Nam, N.D.; Kim, J.G. Effect of niobium on the corrosion behaviour of low alloy steel in sulfuric acid solution. *Corros. Sci.* **2010**, *52*, 3377–3384. [CrossRef]
27. Lee, D.Y.; Kim, W.C.; Kim, J.G. Effect of nitrite concentration on the corrosion behaviour of carbon steel pipelines in synthetic tap water. *Corros. Sci.* **2012**, *64*, 105–114. [CrossRef]
28. Lopez, D.A.; Simison, S.; De Sanchez, S. The influence of steel microstructure on CO_2 corrosion. EIS studies on the inhibition efficiency of benzimidazole. *Electrochim. Acta* **2003**, *48*, 845–854. [CrossRef]
29. Alves, V.A.; Brett, C.M. Characterization of passive films formed on mild steels in bicarbonate solution by EIS. *Electrochim. Acta* **2002**, *47*, 2081–2091. [CrossRef]
30. Brown, R.; Alias, M.; Fontana, R. Effect of composition and thickness on corrosion behavior of TiN and ZrN thin films. *Surf. Coat. Technol.* **1993**, *62*, 467–473. [CrossRef]
31. Hong, M.-S.; Park, I.-J.; Kim, J.-G. Alloying effect of copper concentration on the localized corrosion of aluminum alloy for heat exchanger tube. *Met. Mater. Int.* **2017**, *23*, 708–714. [CrossRef]
32. Bessone, J.; Salinas, D.; Mayer, C.; Ebert, M.; Lorenz, W. An EIS study of aluminium barrier-type oxide films formed in different media. *Electrochim. Acta* **1992**, *37*, 2283–2290. [CrossRef]
33. Mansfeld, F. Electrochemical impedance spectroscopy (EIS) as a new tool for investigating methods of corrosion protection. *Electrochim. Acta* **1990**, *35*, 1533–1544. [CrossRef]
34. Tan, Y.-J.; Bailey, S.; Kinsella, B. An investigation of the formation and destruction of corrosion inhibitor films using electrochemical impedance spectroscopy (EIS). *Corros. Sci.* **1996**, *38*, 1545–1561. [CrossRef]
35. Zeng, A.; Liu, E.; Annergren, I.; Tan, S.; Zhang, S.; Hing, P.; Gao, J. EIS capacitance diagnosis of nanoporosity effect on the corrosion protection of DLC films. *Diam. Relat. Mater.* **2002**, *11*, 160–168. [CrossRef]
36. Cottis, R.; Turgoose, S. *Electrochemical Impedance and Noise*; National Assn of Corrosion Engineers: Houston, TX, USA, 1999.
37. Manov, S.; Lamazouere, A.; Aries, L. Electrochemical study of the corrosion behaviour of zinc treated with a new organic chelating inhibitor. *Corros. Sci.* **2000**, *42*, 1235–1248. [CrossRef]
38. Rondelli, G.; Torricelli, P.; Fini, M.; Rimondini, L.; Giardino, R. In vitro corrosion study by EIS of an equiatomic NiTi alloy and an implant quality AISI 316 stainless steel. *J. Biomed. Mater. Res. B Appl. Biomater.* **2006**, *79*, 320–324. [CrossRef] [PubMed]
39. Hong, M.-S.; Hwang, J.-H.; Kim, J.H. Optimization of the Cathodic Protection Design in Consideration of the Temperature Variation for Offshore Structures. *Corrosion* **2017**, *74*, 123–133. [CrossRef]
40. Bentiss, F.; Jama, C.; Mernari, B.; El Attari, H.; El Kadi, L.; Lebrini, M.; Traisnel, M.; Lagrenée, M. Corrosion control of mild steel using 3,5-bis (4-methoxyphenyl)-4-amino-1,2,4-triazole in normal hydrochloric acid medium. *Corros. Sci.* **2009**, *51*, 1628–1635. [CrossRef]
41. Hong, M.-S.; Kim, S.-H.; Im, S.-Y.; Kim, J.-G. Effect of ascorbic acid on the pitting resistance of 316L stainless steel in synthetic tap water. *Met. Mater. Int.* **2016**, *22*, 621–629. [CrossRef]
42. Hassan, H.H.; Abdelghani, E.; Amin, M.A. Inhibition of mild steel corrosion in hydrochloric acid solution by triazole derivatives: Part I. Polarization and EIS studies. *Electrochim. Acta* **2007**, *52*, 6359–6366. [CrossRef]
43. Liangcai, L.; Ming, W.; Huoming, S.; Haiying, L.; Qingdong, Q.; Yuanlong, D. Preparation and EIS studies on polyimide/polyaniline blend film for corrosion protection. *Polym. Adv. Technol.* **2001**, *12*, 720–723. [CrossRef]
44. Zin, I.; Lyon, S.; Hussain, A. Under-film corrosion of epoxy-coated galvanised steel: An EIS and SVET study of the effect of inhibition at defects. *Prog. Org. Coat.* **2005**, *52*, 126–135. [CrossRef]
45. Marangoni, A.G.; Acevedo, N.; Maleky, F.; Peyronel, F.; Mazzanti, G.; Quinn, B.; Pink, D. Structure and functionality of edible fats. *Soft Matter* **2012**, *8*, 1275–1300. [CrossRef]

46. Acevedo, N.C.; Marangoni, A.G. Nanostructured fat crystal systems. *Annu. Rev. Food Sci. Technol.* **2015**, *6*, 71–96. [CrossRef] [PubMed]
47. Nan, H.; Wang, Z.; Wang, W.; Liang, Z.; Lu, Y.; Chen, Q.; He, D.; Tan, P.; Miao, F.; Wang, X.; et al. Strong photoluminescence enhancement of MoS_2 through defect engineering and oxygen bonding. *ACS Nano* **2014**, *8*, 5738–5745. [CrossRef] [PubMed]
48. Hu, X.; Jiang, P.; Wan, J.; Xu, Y.; Sun, X. Study of corrosion and friction reduction of electroless Ni-P coating with molybdenum disulfide nanoparticles. *J. Coat. Technol. Res.* **2009**, *6*, 275–281. [CrossRef]
49. Rad, N.E.; Dehghanian, C. Effects of co-deposition of Cr_2O_3 and MoS_2 on corrosion properties of nanocomposite electroless nickel coating. *Iran. J. Mater. Sci. Eng.* **2010**, *7*, 1–7.

coatings

MDPI

Article

Delayed Formation of Thermally Grown Oxide in Environmental Barrier Coatings for Non-Oxide Ceramic Matrix Composites

Hagen Klemm *, Katrin Schönfeld and Willy Kunz

Fraunhofer Institute for Ceramic Technologies and Systems IKTS, Dresden 01277, Germany; katrin.schoenfeld@ikts.fraunhofer.de (K.S.); willy.kunz@ikts.fraunhofer.de (W.K.)

* Correspondence: hagen.klemm@ikts.fraunhofer.de; Tel.: +49-351-2553-7553

Received: 22 November 2019; Accepted: 17 December 2019; Published: 19 December 2019

Abstract: The oxidation and corrosion behavior at elevated temperatures of a $SiC_F/SiC(N)$ composite with two plasma-sprayed environmental barrier coating (EBC) systems were studied. After both processes, the formation of a silica-based thermally grown oxide (TGO) layer was observed. The formation of this TGO caused two principal failure mechanisms of the EBC. Firstly, spallation of the EBC induced by stresses from volume expansion and phase transformation to crystalline SiO_2 was observed. Water vapor corrosion of the TGO with gap formation in the top region of the TGO was found to be a second failure mechanism. After a burner rig test of the Al_2O_3-YAG EBC system, this corrosion process was observed at the TGO surface and in the volume of the Al_2O_3 bond coat. In the case of the second system, Si-$Yb_2Si_2O_7$/SiC-Yb_2SiO_5, the formation of the TGO could be delayed by introducing an additional intermediate layer based on $Yb_2Si_2O_7$ filled with SiC particles. The SiC particles in the intermediate layer were oxidized and served as getter to reduce the permeation of oxidants (O_2, H_2O) into the material. In this way, the formation of the TGO and the failure mechanisms caused by their formation and growth could be delayed.

Keywords: environmental barrier coatings; non-oxide ceramic matrix composites; oxidation; water vapor corrosion; thermally grown oxide; damage mechanisms

1. Introduction

Silicon carbide fiber-based ceramic matrix composites (CMCs) offer a high potential for applications as structural components in advanced gas turbines. In comparison to metallic super alloys, used in the state of the art, the main advantages of these materials were found to be their low specific weight in combination with a superior potential at elevated temperatures up to 1400 °C. Furthermore, among ceramic materials, CMCs are characterized by a damage-tolerant fracture behavior, suggesting them as promising candidates for gas turbine applications as well. During recent years, significant progress has been achieved in material development and processing. However, there are still considerable deficits at present, especially in the long-term behavior of the composites in hot gas atmosphere. Corrosion processes were observed, caused by the high water vapor pressure in combination with high temperatures and gas velocities. The resulting microstructural and mechanical degradation of the composites and the damage mechanisms of these processes have been described in several studies [1–6]. Volatilization of the protective silica-based surface layer by the formation and evaporation of silicon hydroxides ($Si(OH)_4$) was found to be the main process leading to considerable material loss with recession rates in the range of 1 µm/h (Equation (1)). Additional material degradation as a consequence of oxidation processes inside the composite was observed.

$$SiO_2 + 2H_2O_{(g)} = Si(OH)_{4(g)}. \quad (1)$$

Environmental barrier coatings (EBCs) have been the solution to prevent the surface corrosion of the ceramic materials in gas turbine atmospheres [7,8]. During the last few years, different EBC systems have been introduced [8–10]. As a consequence of the complex conditions during operation at elevated temperatures in a hot gas atmosphere, multilayer coatings with special functions were proposed. In this way, several features required to guarantee the long-term stability of the EBC system in hot gas conditions could be realized.

Beside their stability against erosion, interaction with Ca-Mg-Al-silicates (CMAS), or foreign object damage, the top layer of the system must primarily exhibit a superior water vapor corrosion stability. During recent years, several oxide systems with superior corrosion stability have been suggested to protect non-oxide ceramics or CMCs based on Si_3N_4 or SiC against water vapor corrosion [11,12]. Among these oxide systems, rare-earth (RE) silicates have been identified as promising EBC candidates for top layer materials. While the RE-monosilicates are mostly stable in a hot gas environment, the disilicates were found to be partially volatilized with the formation of silicon hydroxide and the more stable monosilicate (Equation (2)) [8,12–17].

$$RE_2Si_2O_7 + 2H_2O_{(g)} = RE_2SiO_5 + Si(OH)_{4(g)} \quad (2)$$

This corrosion process led consequently to the formation of a stable monosilicate layer, influencing the corrosion behavior of the EBC system during long-term application.

Currently, a layer based on metallic silicon is used as a very effective bond coat in several EBC systems [18,19]. With a melting point of about 1410 °C, silicon bond coats are limited in their temperature potential. For use at lower temperatures, however, they are characterized by several benefits. First, the coatings agree well with the coefficient of thermal expansion (CTE) of the non-oxide CMC substrate material. The second point is the getter function of the silicon against the permeation of oxidizing compounds (O_2, H_2O) and prevention of oxidation processes inside the CMC component.

Various multilayer EBC systems with Si bond coats and RE-top coats were demonstrated to be quite effective in the protection of SiC_F/SiC CMC [10,16,17,20,21]. However, during operation at elevated temperatures in hot gas atmosphere, several processes led to degradation of the whole system. A summary of possible failure modes was reported by Lee [19]. Processes like the formation of stresses during thermal cycling, foreign objects, phase transformation, or sintering processes resulted in cracking, delamination, and spallation of the EBC system. Additional chemical processes like water vapor or CMAS corrosion limit the stability and functionality of the protecting system. During long-term use, oxidation processes were found to be an additional critical factor for the stability of the EBC. Diffusion of oxygen and, especially, the permeation of H_2O through the different EBC layers are responsible for the formation of a thermally grown oxide (TGO) layer of mainly silica at the upper side of the Si bond coat. With growing thickness of the SiO_2 TGO layer, crystallization and phase transition processes (cristobalite) were observed, finally leading to stresses in the EBC system with the consequence of cracking and spallation of the EBC [10,16,22].

In real conditions, the formation of the TGO layer cannot be avoided. There will always be permeation of oxygen and water into the material, finally leading to oxidation processes inside. However, there are strategies to minimize the rate of TGO layer formation and their following influences. First, the transport of the oxidants (O_2 and H_2O) through the EBC system should be considered. There is still a considerable lack of data about the permeation properties (diffusion coefficient, oxidant solubility) of the various materials used in the EBC layer system. Furthermore, the morphology of the different layers, like layer thickness, porosity, crystallinity, and grain boundary structure, has to be studied regarding their influence on their permeation properties. Recently, an example in this direction was introduced by Lee [19]. The TGO growth rate in an EBC system with the Si bond coat and $Yb_2Si_2O_7$ was found to be significantly reduced by modifying the $Yb_2Si_2O_7$ layer with various oxides (Al_2O_3, mullite, $Y_3Al_5O_{12}$). As a conclusion of these results, he suggested a modification of the SiO_2 network structure of the TGO by incorporating Al^{3+}- and Yb^{3+}-ions, consequently leading to lower permeation rates of oxidants (H_2O) through the TGO layer.

A defined control of the oxidation and corrosion processes itself was found to be a second strategy. This can be performed by modification of the oxidation mechanism, e.g., defined reaction products or the location where the oxidation takes place. An example for such a strategy was reported for monolithic Si_3N_4 with SiC or $MoSi_2$ additions [12,23]. During oxidation of these composites, a changed oxidation mechanism with the formation of Si_2ON_2 in the top region of the bulk material was observed, leading finally to less defect formation caused by the oxidation processes in the microstructure of the material. The focus of this study was placed in a similar direction, namely, to avoid the formation of a TGO as a reaction layer at, e.g., the silicon bond coat, by a defined reaction of the permeating oxidants at other regions.

2. Materials and Methods

The base CMC material was fabricated by winding technology with polycrystalline SiC fibers, Tyranno SA3 (UBE Industries, Tokyo, Japan). Prior to winding, the desized SiC tows were infiltrated with an aqueous slurry composed of SiC powder, Sintec 15 (Saint Gobain, Courbevoie, France), and 20 vol.% sintering additives with Al_2O_3, AKP 50 (Sumitomo Chemical, Tokyo, Japan); Y_2O_3, Grade C (H.C. Stark, Goslar, Germany); and SiO_2, Aerosil Ox 50 (Evonic Industries, Essen, Germany). The wound cylinder (85° winding angle) was cut and pressed, opening into a flat sheet. Matrix formation was performed in five steps of precursor infiltration and pyrolysis (PIP) with commercially available polysilazane Si-C-N precursor, HTT 1800 (Clariant Advanced Materials GmbH, Muttenz, Switzerland). Afterward, the composite was sintered at 1400 °C in nitrogen atmosphere. Finally, a SiC_F/SiC(N) composite with a fiber volume content between 40% and 50% and an open porosity of about 10% was obtained. Further details about the CMC fabrication are described in [24]. Bars with dimensions of $3 \times 10 \times 36\ mm^3$ were used as test samples.

The first EBC system was a bond coat from Al_2O_3 with a top coat of yttrium aluminum garnet ($Y_3Al_5O_{12}$, YAG). Both layers were fabricated by atmospheric plasma spraying. The second system was a three-layer coating system with a Si bond coat, an intermediate layer consisting of a mixture of $Yb_2Si_2O_7$/SiC and Yb_2SiO_5 as the top coat. While the Si bond coat was fabricated by atmospheric plasma spraying (APS), the two rare-earth-containing layers were fabricated by suspension plasma spraying (SPPS). An overview of the coatings fabricated is given in Table 1:

Table 1. Average coating thickness of the layers in the two EBC systems.

	Thickness/µm	Fabrication
Al_2O_3-YAG		
Al_2O_3	50 to 70	APS
YAG	80 to 120	APS
Si-$Yb_2Si_2O_7$/SiC-Yb_2SiO_5		
Si	30 to 50	APS
$Yb_2Si_2O_7$/SiC	100 to 150	SPPS
Yb_2SiO_5	100 to 150	SPPS

Both EBC systems were tested regarding their oxidation resistance at 1200 °C for 100 h in furnace air. Additionally, hot gas corrosion tests were conducted in a high-temperature burner rig at atmospheric pressure [11]. The coated test samples were blown directly by the hot gas in a reactor tube of solid-state sintered SiC with an inner diameter of 30 mm. The hot gas was composed of the combustion products of natural gas in air and additional water vapor. The conditions of the corrosion tests are summarized in Table 2. Further details regarding the burner rig test are described in [11].

Table 2. Burner rig test performed.

Temperature/°C	Flow Speed/m/s	Water Vapor Pressure/bar	Testing Time/h
1200	100	0.2	100

After both tests, the microstructure of the samples was characterized by means of polished cross-sections with field-emission scanning electron microscopy (Ultra 55, Zeiss, Oberkochen, Germany). Information about the composition of the different layers after oxidation and corrosion was obtained by using energy-dispersive X-ray spectroscopy (EDX; ISIS Si (Li) detector).

3. Results

A summary of the weight changes of the two coating systems in comparison to the base CMC substrate without coating observed during oxidation and the burner rig test is summarized in Table 3. A weight gain was observed for all materials investigated after both tests.

Table 3. Weight change after oxidation and burner rig test, 1200 °C, 100 h. Comparison of EBC-coated systems with base ceramic matrix composites (CMC) material without EBC coating.

Material/EBC	Oxidation		Burner Rig Test	
	m/g	m/m/g/g	m/g	m/m/g/g
CMC	0.027	0.009	0.094	0.046
Al_2O_3-YAG	0.048	0.016	0.19	0.064
Si-$Yb_2Si_2O_7$/SiC-Yb_2SiO_5	0.009	0.003	0.045	0.016

The values are the average of the three samples each. In connection with the interpretation of these results, some inaccuracies as a consequence of inhomogeneous EBC layers (thickness, pores, or cracks) should be considered. Furthermore, during the burner rig test, both processes, weight gain and weight loss, were observed. These results cannot be correlated directly with the TGO scale thickness obtained after the tests; however, some general tendencies can be followed as described in the microstructural discussion of the material investigated.

The main results were obtained by comparison of the microstructure of the CMC with different EBCs after oxidation and hot gas corrosion tests both at 1200 °C and 100 h. After oxidation, a weight gain was observed. During the test, oxygen diffused through the EBC layer and finally reacted at the first non-oxide surface in the system (base SiC_F-SiC(N) composite or Si bond coat). Usually, a TGO layer was formed. In the case of the hot gas corrosion test, a second reaction has to be considered as well. Water vapor penetrated through the EBC layer, reacted with the SiO_2 of the TGO layer, and formed volatile $Si(OH)_4$. As a consequence of the high gas speed, the equilibrium of the corrosion reaction in Equation (1) was strongly shifted to the formation of $Si(OH)_4$. In this way, the water vapor corrosion of the TGO became more dominant, resulting in material loss and a gap formation between the silica TGO and the EBC top layer.

This behavior is described in Figure 1 showing the comparison of polished cross-sections of the microstructure after oxidation and corrosion of a model EBC system consisting of a Si bond coat and an Yb_2SiO_5 top coat. While a TGO was formed (A) after oxidation, corrosion processes with the formation of gaps at the Yb_2SiO_5-Si interface (B) were observed.

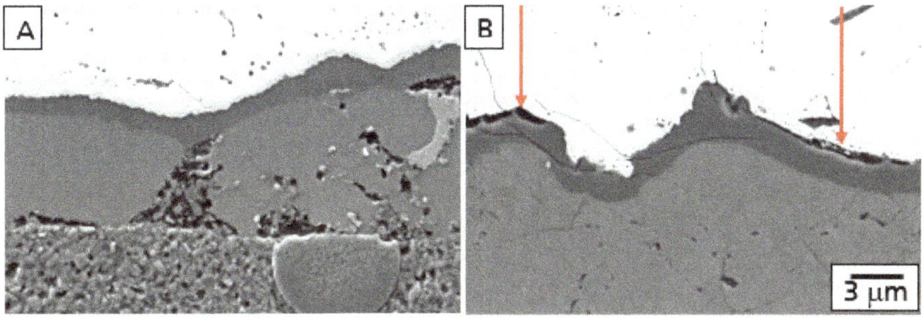

Figure 1. Polished cross-section of TGO in Si-Y_2SiO_5 EBC system after (**A**) 100 h oxidation at 1200 °C in air and (**B**) 100 h hot gas corrosion at 1200 °C.

3.1. CMC Substrate Material

The results of the high-temperature tests of the base CMC without EBC were characterized by investigation of the microstructure of the surface region. After both tests, oxidation and the burner rig, a weight gain was observed. During oxidation, a protective layer of mainly SiO_2 was formed at the surface of the material, limiting the diffusion of oxygen into the material (Figure 2A). This behavior was comparable to dense monolithic SiC. Both oxidation (weight gain) and corrosion (weight loss) were observed during the burner rig test. Caused by the water vapor corrosion of the protecting SiO_2 surface layer, oxygen and water vapor were able to diffuse deeper into the material and oxidized the matrix and the fibers too (Figure 2B). As a consequence of these oxidation processes inside of the material, a higher weight gain was found after the burner rig test.

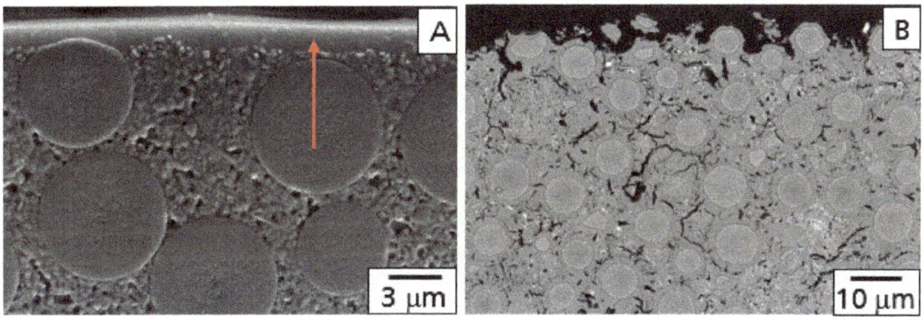

Figure 2. Microstructure of polished cross-sections of the base CMC after thermal treatment. (**A**) 100 h oxidation at 1200 °C; (**B**) 100 h burner rig test at 1200 °C.

3.2. Al_2O_3–YAG EBC Coating System

In the case of the material coated with Al_2O_3-YAG, a weight gain was observed after both tests. The values were found to be higher in comparison to the base CMC. Two effects are assumed to be the reason; first, a low protective ability caused by the inhomogeneity of the double layer with a high amount of cracks and porosity, and second, the TGO as the rate controlling factor for oxidation based on an alumosilicate glass with a significantly higher diffusion ability in comparison to the surface oxidation layer formed during oxidation of the base CMC [25].

The microstructure of the Al_2O_3-YAG EBC coating on the SiC_F-SiC(N) composite in the coated condition is illustrated in Figure 3.

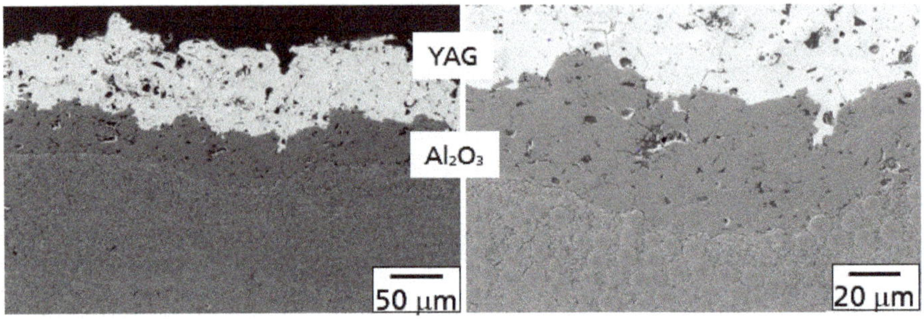

Figure 3. Microstructure of EBC consisting of Al_2O_3 bond coat and YAG top coat.

In principle, both processes, as described above, were observed after oxidation and the hot gas test. A comparison of polished cross-sections with the Al_2O_3/YAG coating is shown in Figure 4 after oxidation in air (A) and the burner rig test (B) at 1200 °C and 100 h.

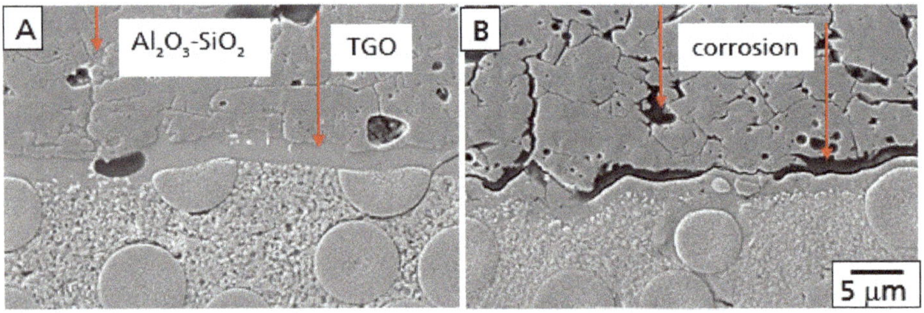

Figure 4. Comparison of microstructure in polished cross-sections of SiC_F/SiC(N) composite coated with Al_2O_3-YAG. (**A**) TGO formation and Al_2O_3-SiO_2 glass in and under the alumina layer after 100 h oxidation at 1200 °C. (**B**) Corrosion of TGO and grain boundary in the volume of Al_2O_3 layer after burner rig test 100 h and 1200 °C.

During the oxidation test, oxygen diffused through the YAG/Al_2O_3 layer and oxidized the SiC fibers and SiC(N)-matrix to SiO_2. Consequently, a TGO layer at the interface of the CMC base material and the Al_2O_3 bond coat was formed. Furthermore, a part of the oxidation product was found in the grain boundaries and triple junctions of the Al_2O_3 bond coat. The composition of the oxidation products in both the TGO and Al_2O_3 layer was a glassy alumosilicate (Figure 4A).

As described above, corrosion processes were observed additionally after the burner rig test (Figure 4B). The alumosilicate glass in both the TGO and the grain boundaries and triple junctions were found to be corroded, forming small pores and voids in the Al_2O_3 bond coat and gaps between the top of the TGO and the bond coat. With increasing time, both corrosion processes will damage the EBC system. Especially, the corrosion of the TGO will form large defects, finally leading to failure of the EBC. The smaller pores and voids in the alumina bond coat, however, are much more stable from a mechanical point of view.

Notwithstanding other damage mechanisms, this behavior opens an idea to the enhance high-temperature stability of the EBC in principle. By the shifting of the oxidation processes from the interface of the TGO into the volume of the bond coat, it should be possible to decelerate the TGO formation. Furthermore, the damage mechanism in the TGO (crystallization processes and corrosion) could be avoided, and only smaller corrosion defects in the bond coat should be observed.

3.3. Si–$Yb_2Si_2O_7$/SiC–Yb_2SiO_5 EBC Coating System

This mechanism described above was considered in the design of a three-layer coating system investigated next with a silicon bond coat, an intermediate layer consisting of $Yb_2Si_2O_7$ with SiC particles, and Yb_2SiO_5 as the top coat featured by a superior hot gas stability. The microstructure of this EBC is demonstrated in Figure 5. Few microcracks were observed in the Yb_2SiO_5 top layer, probably as a consequence of the CTE mismatch between the top and intermediate layer (CTE $Yb_2Si_2O_7$ 4.2×10^{-6} K^{-1}; CTE Yb_2SiO_5 6.8×10^{-6} K^{-1}).

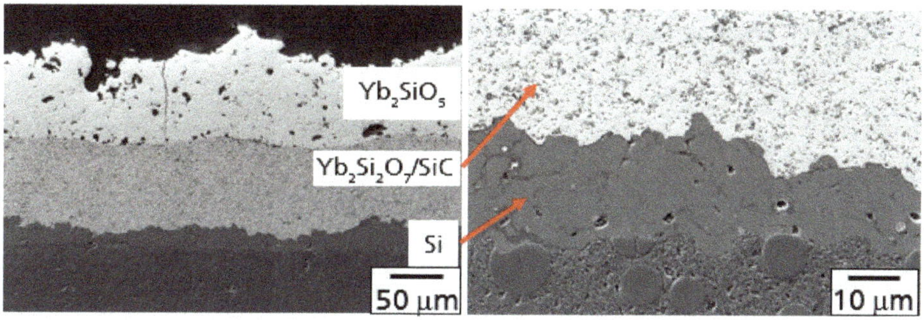

Figure 5. Polished cross-section of EBC layer system with Si bond coat, $Yb_2Si_2O_7$/SiC, and Yb_2SiO_5 layer fabricated by atmospheric plasma spaying (Si) and suspension plasma spraying ($Yb_2Si_2O_7$/SiC, Yb_2SiO_5).

In comparison to the other materials investigated, only a small weight gain was measured after both tests at elevated temperatures. This should be caused by a protecting function, especially of the $Yb_2Si_2O_7$/SiC layer. Oxygen diffusion through the EBC layers and the oxidation reaction in the $Yb_2Si_2O_7$/SiC layer were found to be the main processes observed after the oxidation test. The SiC particles in the intermediate layer were oxidized, consequently leading to the formation of a SiO_2 scale on the SiC particles. The diffusion of oxygen through this SiO_2-based layer controlled the oxidation process of the whole system during the testing time performed. A typical example for the mechanism is demonstrated in Figure 6A, presenting a polished cross-section at the interface of the Yb_2SiO_5 top coat and the $Yb_2Si_2O_7$/SiC intermediate layer.

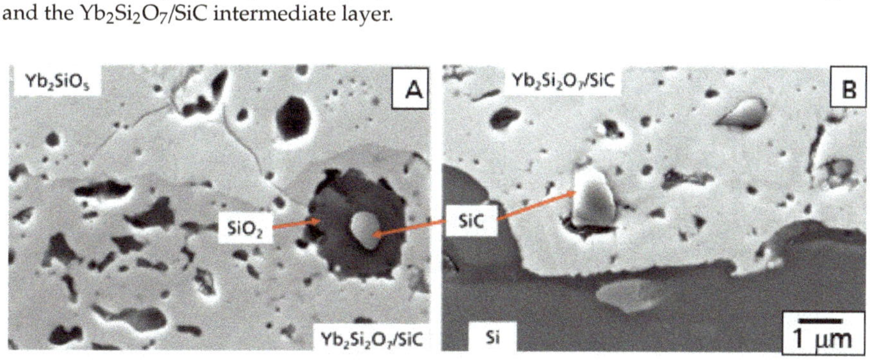

Figure 6. Polished cross-section of Si–$Yb_2Si_2O_7$/SiC–Yb_2SiO_5 EBC coating system. SiC particle in $Yb_2Si_2O_7$ layer at interfaces (**A**) Yb_2SiO_5–$Yb_2Si_2O_7$/SiC and (**B**) $Yb_2Si_2O_7$/SiC–Si after oxidation for 100 h and 1200 °C.

The SiC particles in this layer operated as a getter for the diffusion of oxygen, preventing further oxygen at diffusing deeper into the material. This effect is illustrated in Figure 6B with the microstructure at the interface between the Si bond coat and the $Yb_2Si_2O_7$/SiC intermediate layer. As the diffusion of oxygen reacted with SiC particles in the upper region of this layer, no oxidation processes were observed in this area. The SiC grains did not show an oxidation layer. Furthermore, the formation of a TGO was not observed. Similar results on the graded oxidation of SiC-particles have been reported in context of the mechanical self-healing ability of $Yb_2Si_2O_7$/SiC composites [26,27].

In Figure 7, the microstructure of this EBC system is shown after the 100 h hot gas corrosion test at 1200 °C. Although some additional cracks formed, the EBC coating is still intact, protecting the CMC from the direct hot gas corrosion attack (Figure 7A).

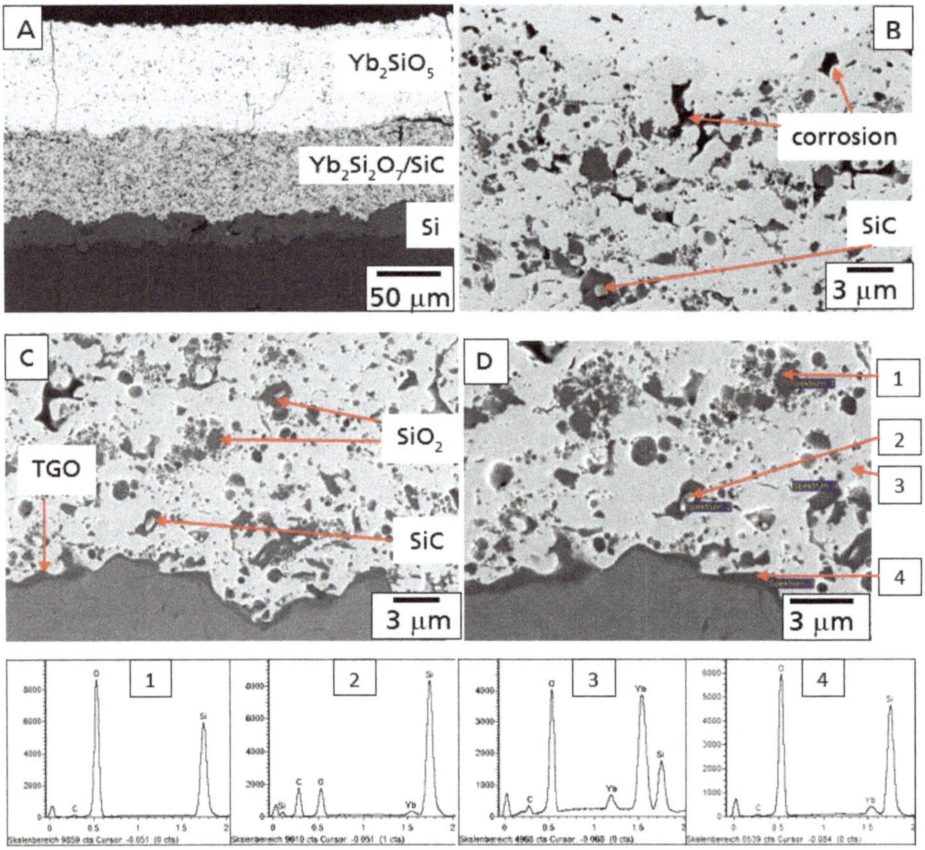

Figure 7. Polished cross-section of SiC_F/SiC(N) composite with three-layer EBC (Si–$Yb_2Si_2O_2$/SiC–Yb_2SiO_5) after 100 h hot gas test (**A**) at 1200 °C, overview; (**B**) at the interface to the Yb_2SiO_5 top coat and the $Yb_2Si_2O_7$/SiC interlayer, and (**C**) at the interface between the $Yb_2Si_2O_7$/SiC interlayer and the Si bond coat; energy-dispersive X-ray spectroscopy (EDX) analysis of different phases in the microstructure, analysis locations demonstrated in (**D**).

In comparison to the oxidation test, stronger oxidation processes were observed in the $Yb_2Si_2O_7$/SiC layer after the test in hot gas conditions. A higher amount of oxidants reached and oxidized the SiC particles in the intermediate layer of the EBC coating system as a consequence of additional water vapor permeation. In agreement to the literature [28,29], water vapor was found to be the dominant oxidant caused by the significantly higher solubility of H_2O in SiO_2. A second reasonable possibility is the formation of interconnected open splat boundaries in the Yb-silicate layer [26]. Further investigation has to be performed in this field. The SiC particles were found to be oxidized to SiO_2 up to the bottom of the $Yb_2Si_2O_7$/SiC layer. As a part of the oxidant reached the Si-bond coat the TGO layer started to grow (Figure 7C). Results of EDX analysis of microstructural details of this part are presented in Figure 7.

Additional processes were observed in the top region of the $Yb_2Si_2O_7$/SiC layer. Some of the SiO_2 areas formed by the oxidation of the SiC particles were found to be corroded by water vapor, finally leading to the formation of pores in the top region of the intermediate layer (Figure 7B).

4. Discussion

As already mentioned above, the penetration of oxidants into the EBC system and the following formation of a TGO cannot be prevented completely. However, with purposeful design of the EBC layer system, the transport processes and the following oxidation processes can be influenced. The first possibility should be the reduction of the transport processes for O_2 and H_2O. Coatings with a lower oxygen diffusion coefficient and water vapor solubility are reasonable trends for material development.

In the present study, the following oxidation and corrosion processes were modified. With the introduction of oxidable particles into an intermediate layer, it was possible to shift the oxidation processes from the interface between the Si bond coat and the Yb-silicate top coat to a volume in the intermediate layer. The processes occurring during oxidation and corrosion are schematically described in Figure 8.

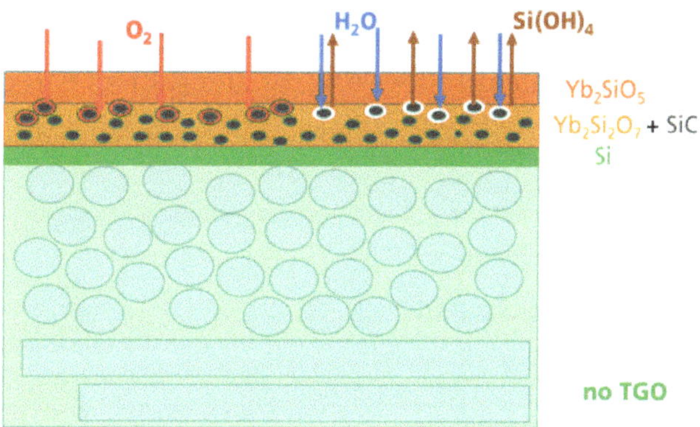

Figure 8. Schematic diagram of the oxidation and corrosion mechanism of SiC_F/SiC(N) composite with three-layer EBC system.

Independent of the thermal treatment, oxidation, or hot gas corrosion, the transport of oxidants and the following oxidation processes in the EBC system were found to be the first processes in the system. The penetrated O_2 and H_2O reacted with the SiC, resulting in the formation of a SiO_2-based shell surrounding the SiC particles. In this way, the SiC particles served as a getter for the further transport of oxidants into deeper regions of the whole system. As the oxidants could not reach the Si bond coat, the formation of the TGO was prevented. In a hot gas atmosphere with increased water vapor pressure and high gas velocity, corrosion processes were observed. The water vapor penetrated this region and reacted with the SiO_2 at the surface of the SiC particles and formed $Si(OH)_4$. A $Si(OH)_4$ gradient developed as a consequence of the high hot gas velocity outside, which was a driving force of the outward transport and evaporation of the $Si(OH)_4$. Small pores were found to be the result of the corrosion process.

The beneficial gettering function of the SiC particulate will be a temporary effect only. After longer oxidation time or at higher temperatures, the SiC particles will be consumed and oxygen will reach the silicon bond coat to form the TGO. However, with the incorporation of the SiC particles, to be oxidized during operation in hot gas atmosphere, the EBC system could be temporary stabilized. With the delayed formation of the TGO, their resulting damage mechanisms, cracking as a result of stresses by crystallization and phase transition processes and the gap formation caused by corrosion material loss, started at later application times. A very beneficial effect in terms of long-term stability and lifetime can only be achieved by simultaneous improvement of the oxygen and H_2O permeation behavior of the whole EBC layer system. The lower the transport of the oxidants into the material, the longer the gettering function of the SiC particles can be used. Further studies have to be performed to optimize the EBC system regarding composition and microstructure with special focus on the transport mechanisms during service in hot gas environments.

Author Contributions: H.K. conceptualized the idea; H.K., K.S. and W.K. developed the coating; K.S. and W.K. performed the experiments and analyzed data; H.K. supervised the project and acquired funding; H.K. prepared the manuscript; K.S. and W.K. contributed in editing and submission. All authors have read and agreed to the published version of the manuscript.

Funding: This research was funded by the German Federal Ministry for Education and Research, grant number 03EK3544C and Fraunhofer Funding MAVO CMC engine.

Acknowledgments: The authors gratefully thank B. Gronde and F.L. Toma for support in plasma spraying technology.

Conflicts of Interest: The authors declare no conflict of interest.

References

1. Endo, Y.; Tsuchiya, T.; Furuse, Y. Corrosion behavior of ceramics for gas turbines application, silicon based structural ceramics. *Ceram. Trans.* **1994**, *42*, 319–326.
2. Opila, E.J.; Hann, R.E. Paralinear Oxidation of CVD SiC in water vapor. *J. Am. Ceram. Soc.* **1997**, *80*, 197–205. [CrossRef]
3. Yuri, I.; Hisamatsu, T. Recession Rate Prediction for Ceramic Materials in Combustion Gas Flow. In Proceedings of the ASME Turbo Expo 2003, Atlanta, GA, USA, 16–19 June 2003.
4. Filsinger, D.; Schulz, A.; Wittig, S.; Taut, C.; Klemm, H.; Wötting, G. Model Combustor to Assess the Oxidation Behavior of Ceramic Materials under Real Engine Conditions. In Proceedings of the ASME Turbo Expo '99, Indianapolis, Indianapolis, IN, USA, 7–10 June 1999.
5. More, L.K.; Tortorelli, P.F.; Ferber, M.K.; Keiser, L.R. Observations of accelerated silicon carbide recession by oxidation at high water-vapor pressures. *J. Am. Ceram. Soc.* **2000**, *83*, 211–223. [CrossRef]
6. Klemm, H.; Schubert, C.; Taut, C.; Schulz, A.; Wötting, G. Corrosion of Non-Oxide Silicon-Based Ceramics in a Gas Turbine Environment. In Proceedings of the 7th Symposium on Ceramic Materials & Components for Engines, 2000, Weinheim, Germany, 29 March 2001; WILEY-VCH Verlag: Weinheim, Germany, 2001; pp. 153–156.
7. Lee, K.N.; Fritze, H.; Ogura, Y. Coatings for engineering ceramics. In *Progress in Ceramic Gas Turbine Development*; van Roode, M., Ferber, M., Richerson, D.W., Eds.; ASME Press: New York, NY, USA, 2003; Volume 2, pp. 641–664.
8. Lee, K.N. Environmental barrier coatings for CMCs. In *Ceramic Matrix Composites*; Bansal, N.P., Lamon, J., Eds.; Wiley: New York, NY, USA, 2015; pp. 430–451.
9. Lee, K.N. Current status of environmental barrier coatings for Si-based ceramics. *Surf. Coat. Technol.* **2000**, *133*, 1–7. [CrossRef]
10. Lee, K.N.; Fox, D.S.; Bansal, N.P. Rare earth silicate environmental barrier coatings for SiC/SiC composites and Si3N4 ceramics. *J. Eur. Ceram. Soc.* **2005**, *25*, 1705–1715. [CrossRef]
11. Fritsch, M. Heißgaskorrosion keramischer Werkstoffe in H2O-haltigen Rauchgasatmosphären. In *Dissertationsschrift, TU Dresden*; Fraunhofer IRB Verlag: Stuttgart, Germany, 2007; ISBN 978-3-8167-7588-1.
12. Klemm, H. Silicon nitride for high-temperature applications. *J. Am. Ceram. Soc.* **2010**, *93*, 1501–1522. [CrossRef]
13. Ueno, S.; Ohji, T.; Lin, H.T. Recession behavior of Yb2Si2O7 phase under high speed steam jet at high temperatures. *Corros. Sci.* **2008**, *50*, 178–182. [CrossRef]

14. Yuri, I.; Hisamatsu, T.; Ueno, S.; Ohji, T. Exposure test results of Lu2Si2O7 in combustion gas flow at high temperature and high speed. *Jpn. Soc. Mech. Eng.* **2001**, *44*, 520–527.
15. Hong, Z.; Cheng, L.; Zhang, L.; Wang, Y. Water vapor corrosion behavior of scandium silicates at 1400 °C. *J. Am. Ceram. Soc.* **2009**, *92*, 193–196. [CrossRef]
16. Richards, B.T.; Young, K.A.; de Francqueville, F.; Sehr, S.; Begley, M.R.; Wadley, H.N.G. Response of ytterbium disilicate–silicon environmental barrier coatings to thermal cycling in water vapor. *Acta. Mater.* **2016**, *106*, 1–14. [CrossRef]
17. Bakan, E.; Sohn, Y.J.; Kunz, W.; Klemm, H.; Vaßen, R. Effect of processing on high-velocity water vapor recession behavior of Yb-silicate environmental barrier coatings. *J. Europ. Ceram. Soc.* **2019**, *39*, 1507–1513. [CrossRef]
18. Lee, K.N.; Fox, D.S.; Eldridge, J.I.; Zhu, D.; Robinson, R.C.; Bansal, N.P.; Miller, R.A. Upper temperature limit of environmental barrier coatings based on mullite and BSAS. *J. Am. Ceram. Soc.* **2003**, *86*, 1299–1306. [CrossRef]
19. Lee, K.N. Yb2Si2O7 Environmental barrier coatings with reduced bond coat oxidation rates via chemical modifications for long life. *J. Am. Ceram. Soc.* **2019**, *102*, 1507–1521. [CrossRef]
20. Xu, Y.; Hu, X.; Xu, F.; Li, K. Rare earth silicate environmental barrier coatings: Present status and prospective. *Ceram. Int.* **2017**, *43*, 5847–5855. [CrossRef]
21. Zhong, X.; Niu, Y.; Li, H.; Zhou, H.; Dong, S.; Zheng, X.; Ding, C.; Sun, J. Thermal shock resistance of tri-layer Yb2SiO5/Yb2Si2O7/Si coating for SiC and SiC-matrix composites. *J. Am. Ceram. Soc.* **2018**, *101*, 4743–4752. [CrossRef]
22. Lu, Y.; Luo, L.; Liu, J.; Zhu, C.; Wang, Y. Failure mechanism associated with the thermally grown silica scale in environmental barrier coated C/SiC composites. *J. Am. Ceram. Soc.* **2016**, *99*, 2713–2719. [CrossRef]
23. Klemm, H.; Herrmann, M.; Schubert, C. High temperature oxidation and corrosion of silicon-based nonoxide ceramics. *J. Eng. Gas. Turbines Power* **2000**, *122*, 13–18. [CrossRef]
24. Schönfeld, K.; Klemm, H. Interaction of fiber matrix bonding in SiC/SiC ceramic matrix composites. *J. Europ. Ceram. Soc.* **2019**, *39*, 3557–3565. [CrossRef]
25. Courtright, E.L. Engineering property limitations of structural ceramics and ceramic composites above 1600 °C. *Ceram. Eng. Sci. Proc.* **1991**, *12*, 1725–1744.
26. Kunz, W. Entwicklung selbstheilender Materialien zum Korrosionsschutz für Keramiken in Hochtemperaturanwendungen. In *Dissertationsschrift, TU Dresden*; Fraunhofer: Munich, Germany, 2019; ISBN 978-3-8396-1507-2.
27. Kunz, W.; Klemm, H. Self-healing EBC material for gas turbine applications. Advances in high temperature ceramic matrix composites and materials for sustainable development. *Ceram. Trans.* **2018**, *263*, 173–185.
28. Deal, B.E.; Grove, A.S. General relationship for the thermal oxidation of silicon. *J. Appl. Phys.* **1965**, *36*, 3770–3778. [CrossRef]
29. Opila, E.J. Variation of the oxidation rate of silicon carbide with water-vapor pressure. *J. Am. Ceram. Soc.* **1999**, *82*, 625–636. [CrossRef]

coatings

MDPI

Article

Cavitation Erosion and Sliding Wear Mechanisms of AlTiN and TiAlN Films Deposited on Stainless Steel Substrate

Mirosław Szala [1,*], Mariusz Walczak [1], Kamil Pasierbiewicz [1,2] and Mariusz Kamiński [3]

[1] Department of Materials Engineering, Faculty of Mechanical Engineering, Lublin University of Technology, Nadbystrzycka 36D, 20-618 Lublin, Poland; m.walczak@pollub.pl (M.W.); pasierbiewicz.k@gmail.com (K.P.)
[2] Faculty of Transport and Computer Science, University of Economics and Innovation, Projektowa 4, 20-209 Lublin, Poland
[3] Department of Automotive Vehicles, Faculty of Mechanical Engineering, Lublin University of Technology, Nadbystrzycka 36D, 20-618 Lublin, Poland; mariusz.kaminski@pollub.pl
* Correspondence: m.szala@pollub.pl; Tel.: +48-815-384-209

Received: 18 April 2019; Accepted: 23 May 2019; Published: 25 May 2019

Abstract: The resistance to cavitation erosion and sliding wear of stainless steel grade AISI 304 can be improved by using physical vapor deposited (PVD) coatings. The aim of this study was to investigate the cavitation erosion and sliding wear mechanisms of magnetron-sputtered AlTiN and TiAlN films deposited with different contents of chemical elements onto a stainless steel SS304 substrate. The surface morphology and structure of samples were examined by optical profilometry, light optical microscopy (LOM) and scanning electron microscopy (SEM-EDS). Mechanical properties (hardness, elastic modulus) were tested using a nanoindentation tester. Adhesion of the deposited coatings was determined by the scratch test and Rockwell adhesion tests. Cavitation erosion tests were performed according to ASTM G32 (vibratory apparatus) in compliance with the stationary specimen procedure. Sliding wear tests were conducted with the use of a nano-tribo tester, i.e., ball-on-disc apparatus. Results demonstrate that the cavitation erosion mechanism of the TiAlN and AlTiN coatings rely on embrittlement, which can be attributed to fatigue processes causing film rupture and internal decohesion in flake spallation, and thus leading to coating detachment and substrate exposition. At moderate loads, the sliding wear of thin films takes the form of grooving, micro-scratching, micro-ploughing and smearing of the columnar grain top hills. Compared to the SS reference sample, the PVD films exhibit superior resistance to sliding wear and cavitation erosion.

Keywords: cavitation erosion; sliding wear; thin film; AlTiN; TiAlN; mechanical properties; stainless steel; magnetron sputtering; wear mechanism

1. Introduction

Stainless steel grade AISI 304 (SS) is one of the most widely used modern structural materials due to its high corrosion resistance, satisfactory mechanical properties, sufficient weldability and good formidability [1–6]. However, its sliding wear and cavitation erosion resistance (CER) is limited. The wear resistance of SS can be improved by depositing different systems of hard coatings, such as TiN, CrN, TiAlN, AlTiN or AlCrN, or even by depositing stainless steel coatings enriched with silver [3,7–10]. Modern types of TiAlN or AlTiN hard films are widely used for manufacturing machining tools or machine components to reduce their tribochemical and adhesive wear, or to make them resistant to severe thermal conditions. In other words, the use of AlTiN and TiAlN coatings (with different ratios of Al/(Ti + Al)) can be an effective way of increasing the wear resistance of stainless steel components exposed to severe wear processes [9–15]. Moreover, universal ternary coating systems such as TiAlN or AlTiN can not only improve the tribological properties of a substrate, but can also be used to create a protective layer on steel components subjected to cavitation erosion, and thus enhance their cavitation erosion resistance.

A survey of the literature on the cavitation erosion (CE) of binary coating systems demonstrates that hard thin films such as TiN, CrN have a higher CER than the SS304 substrate [7,16]. In turn, ternary film systems usually exhibit superior mechanical properties to binary coating systems. Also, it is stated in the literature [7,17] that, besides film properties, mechanical properties of the substrate are crucial for CER and physical vapor deposited (PVD) coatings too. In addition, there are studies investigating the CE of aluminum alloy coated with a thin TiAlN film [17]. However, there are no studies on the cavitation erosion mechanism of TiAlN or AlTiN in relation to the properties of the SS304 substrate. Also, there is a limited number of studies comparing the properties of TiAlN or AlTiN coatings in relation to their cavitation erosion resistance and investigating their CE wear mechanisms. Additionally, it is worth investigating the cavitation wear resistance of TiAlN and AlTiN in order to determine whether the chemical composition differences (i.e., Al/(Ti + Al) ratio) have any effect on their CER.

The nanoindentation results (i.e., hardness and elastic modulus) reported in the literature primarily concern samples before cavitation testing [7,17]. In this work, the relationship between the evolution of mechanical properties of a coating and CE testing is investigated to estimate the cavitation erosion wear mechanism of the magnetron-sputtered films. This is a quite novel approach to film cavitation erosion evaluation.

TiAlN and AlTiN films can easily be applied on various substrates, and are widely used in machine building for preventing from tribological wear. These coatings are known to operate well under high load conditions [9–14]. However, their sliding wear mechanism under moderate load conditions has not been sufficiently reported in the literature.

The aim of this work is to investigate the cavitation erosion and sliding wear under moderate load of magnetron-sputtered AlTiN and TiAlN coatings deposited on a stainless steel substrate. The paper is an introduction to the CER testing of films deposited on various metal alloy substrates and attempt at quantitative determination of their CER. It is proposed that thin films such as TiAlN or AlTiN be applied in fluid machinery, precise mechanics and engine manufacturing as a method for preventing cavitation wear.

2. Materials and Methods

AlTiN and TiAlN films were deposited with different contents of chemical elements on a stainless steel substrate (grade AISI 304) by using DC magnetron sputtering to obtain approx. 3 μm thick layers. The substrate roughness was lower than $Sa = 0.2$ μm. After deposition, several properties were identified. Thickness of the coatings was examined with a Calotester (surface testing), and metallographic examination on the sample sections was performed by light optical microscopy (LOM, Nikon MA200, Nikon Corporation) (Tokyo, Japan), Scanning electron microscopy (SEM, Quanta, Thermo Fisher Scientific—FEI, and Phenom World ProX, BSE, topographic modes and EDS, Phenom World) (Waltham, MA, USA) and an optical profiler (ContourGT-X 3D Optical Profiler, Bruker Corporation, Billerica, MA, USA) were employed to examine structure, surface morphology, chemical composition and roughness of the coatings. The deposited nitride films and mechanical properties of the substrate were tested on the top sample surfaces using an Ultra Nanoindentation Tester (Anton Paar GmbH, Ostfildern, Germany), in compliance with the procedures described in [18]. Hardness H and elastic modulus E were calculated from the indentation load-displacement data by the Oliver Pharr method [19,20]. In accordance with the literature of the subject [14,21–23], the indentation depth was maintained below 10% of the coating thickness. It should be noted that, out of approx. Out of 80 indentations made, only approx. 80%–90% of the results were considered to estimate mechanical properties of the films. Adhesion of the deposited coatings was determined by scratch testing and Rockwell adhesion tests. The scratch tests were performed on a Micro Combi Tester (Anton Paar

GmbH, Ostfildern, Germany) [24] while the Rockwell adhesion tests [15,25] were conducted according to the VDI 3198 procedure.

Cavitation erosion and sliding wear tests were conducted on bare stainless steel grade SS304 (SS) and SS samples coated with PVD films. Cavitation erosion tests were performed according to ASTM G32 [26] (vibratory apparatus), in compliance with the stationary specimen procedure. The test procedure was described in previous work [27], while a schematic design of the apparatus is shown in Figure 1. The specimen is mounted in the holder and submerged in distilled water. Horn tip vibrations generate a cavitation field that causes erosion of the specimen surface (Figure 1). In the present study, a sonotrode tip with a 15.9 mm diameter was used. Damaged samples were examined by weighing with a 0.1 mg accuracy balance. Worn surface roughness was measured with an optical profilometer and examined by the SEM-EDS method and optical microscopy (OM, stereoscope microscope Nikon SMZ 1500) (Tokyo, Japan). The cavitation-worn area was measured with the ImagePro computer image analysis software (Media Cybernetics, Inc., Rockville, MD, USA) and captured as OM images [28]. During 4.5 h of cavitation, normalized cavitation erosion resistance was calculated by dividing the estimated indicators of SS304 by estimated indicators of the AlTiN and TiAlN films, based on the results of sample weighing (mass loss), computer image analysis of worn surfaces (% of the damaged area) and surface roughness (roughness parameter *Sa*). Additionally, a new concept for describing the CE effect on the deposited films was proposed, i.e., nanoindentation measurement of the H and E parameters of the sample surface before and after cavitation erosion testing. In this way, the observed changes in the film properties could be used for CER mechanism description.

To simulate moderate friction conditions describing mostly limited sliding loads in fluid machinery, the nano-tribo tests were performed with the use of the NTR2 nano-tribometer from Anton Paar GmbH [29,30]. A WC ball with a 0.5 mm diameter was used as the counterpart. A load of 0.8 N, a sliding distance of 90 m and a sliding radius of 5 mm were applied to the ball. As a result, the friction coefficient μ of individual samples could be determined. Next, the wear trace of individual samples was measured using the Taylor Hobson Form Talysurf Intra profile measurement gauge. The measurements provided profilograms enabling determination of the wear factor *K*, see [31]. In addition, obtained wear traces were analyzed with SEM to identify the mechanism of wear.

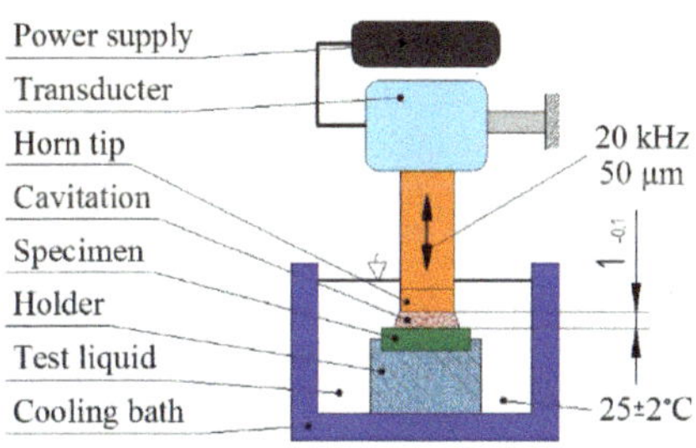

Figure 1. Design of a vibratory apparatus for cavitation tests.

3. Results and Discussion

3.1. Description of Film Properties in Relation to Stainless Steel Substrate

The wear resistance of thin films should be discussed in relation to the structure and properties of the film and substrate alike. In the present work, the thickness of the magnetron-sputtered films measured with a Calotester agrees with the findings of cross-section metallographic examination, amounting to ≈2.7 µm for TiAlN and ≈3.8 µm for AlTiN. The coatings show a dense compact structure (Figure 2) characterized by columnar growth which according to Kalss [32]—is typical of the sputtering technique. The coatings' roughness measured with a profilometer was lower than 0.2 µm, which corresponds to surface preparation of the stainless steel substrate.

This preliminary study investigates the influence of differences in the chemical composition of AlTiN and TiAlN films on the films wear behavior. Chemical compositions of the films are compared in Table 1, showing that the AlTiN film contains more aluminum and nitrogen than the TiAlN film. Both the Al content and Al/(Ti + Al) ratio play an important role in increasing film mechanical properties [9,14]. Therefore, the higher average film-to-SS substrate adhesion of AlTiN (Table 2) can be explained by a higher Al content. It can be claimed that the differences in the chemical composition of AlTiN and TiAlN affect the mechanical properties of these films, as well as the substrate–to-film performance and, finally, functional properties such as wear resistance.

Moreover, a statistical analysis of the Lc2 values and scratch trace (Figures 3 and 4) reveals a satisfactory degree of adhesion for both films. In Figure 3 conformal cracking, cohesive failure and microcracks are visible, while from Figure 4 it can be concluded that the main adhesive failure modes are cohesive failure, spallation and tensile micro-cracking followed by chipping failure of the AlTiN and TiAlN films, which agrees with the results reported for films deposited on SS substrate [12]. Additionally, Figure 4 shows finer cracking of the TiAlN film than the AlTiN, which may result from variations in the coating's elasticity modulus, film adhesion and cohesion.

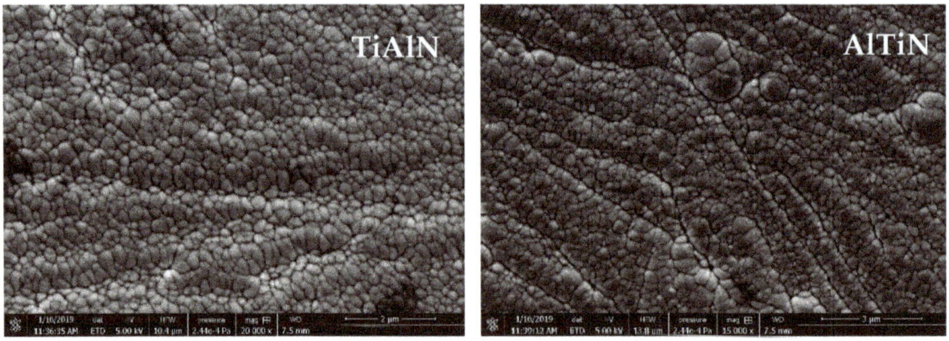

(a) (b)

Figure 2. Surface of the PVD coatings: (**a**) TiAlN; (**b**) AlTiN, SEM.

Table 1. Results of SEM-EDS surface chemical composition spot analysis.

Film	Spot	Chemical Element						
		Ti (wt%)	Al (wt%)	N (wt%)	Ti (at%)	Al (at%)	N (at%)	Al/(Ti + Al)
TiAlN	1	52.93	30.63	16.43	32.39	33.25	34.36	0.507
	2	52.27	30.07	17.67	31.49	32.14	36.38	0.505
	3	50.20	31.10	18.70	29.66	32.59	37.75	0.524
	Average	51.80	30.60	17.60	31.16	32.66	36.18	0.512
AlTiN	1	41.93	35.63	22.43	23.06	34.77	42.17	0.601
	2	43.30	35.10	21.60	24.14	34.71	41.15	0.590
	3	44.97	35.47	19.57	25.73	36.00	38.27	0.583
	Average	43.40	35.40	21.20	24.29	35.15	40.55	0.591

Table 2. Film critical loads estimated by scratch test (Figure 3): Lc1—first symptoms of cohesive failure (angular or parallel cracking), Lc2— beginning of adhesive failure (buckling, chipping, spalling, etc.), Lc3—total failure of the coating or massive exposure of the substrate (mean ± SD).

Film	Lc1 [N]	Lc2 [N]	Lc3 [N]
TiAlN	0.91 ± 0.54	7.23 ± 1.19	15.38 ± 3.43
AlTiN	0.91 ± 0.61	8.87 ± 2.40	18.93 ± 4.76

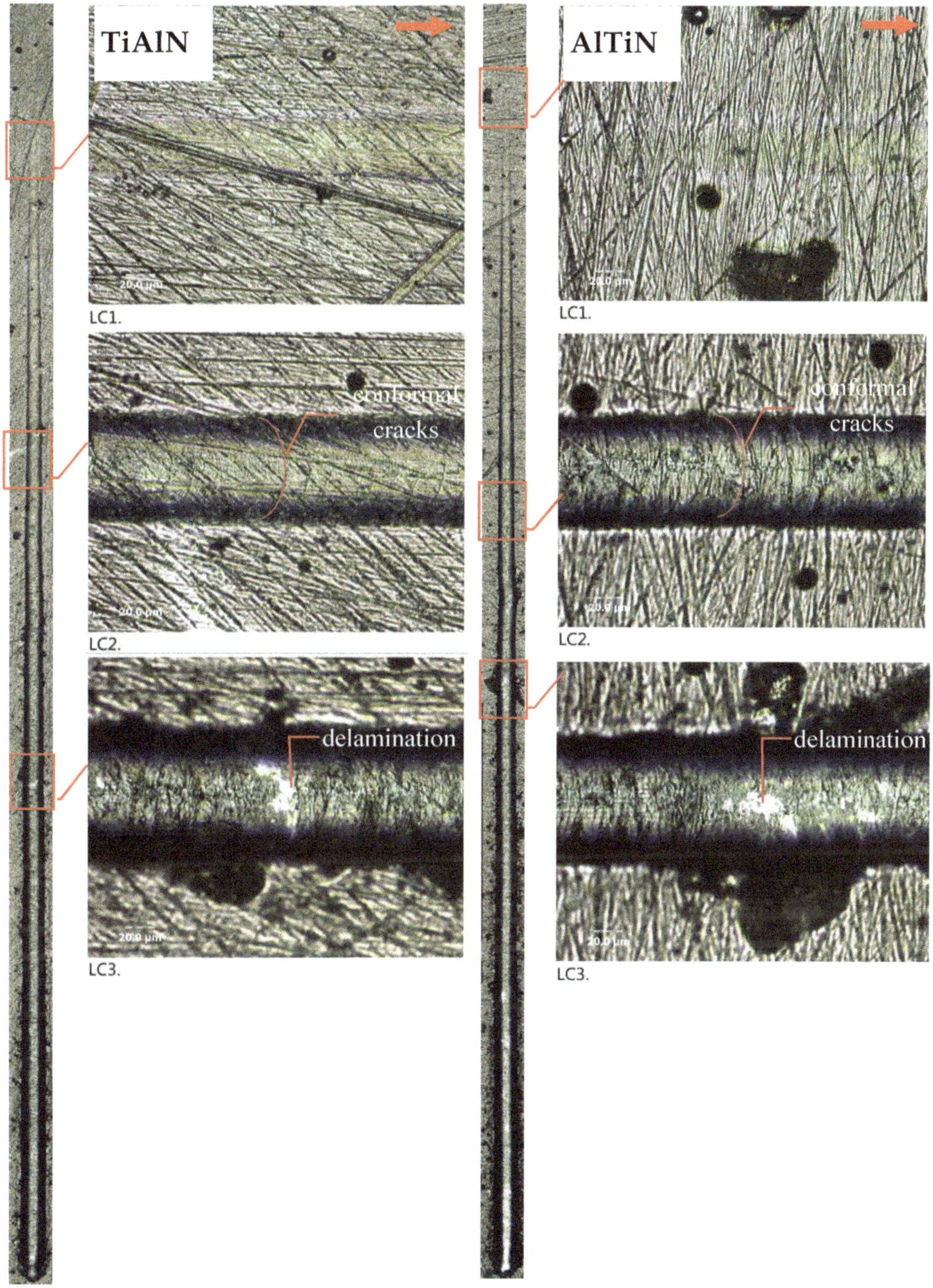

(a) (b)

Figure 3. Scratch traces: (**a**) TiAlN; (**b**) AlTiN and enlarged characteristic areas of Lc1, Lc2 and Lc3. (total scratch trace 3 mm), scratch direction marked by an arrow.

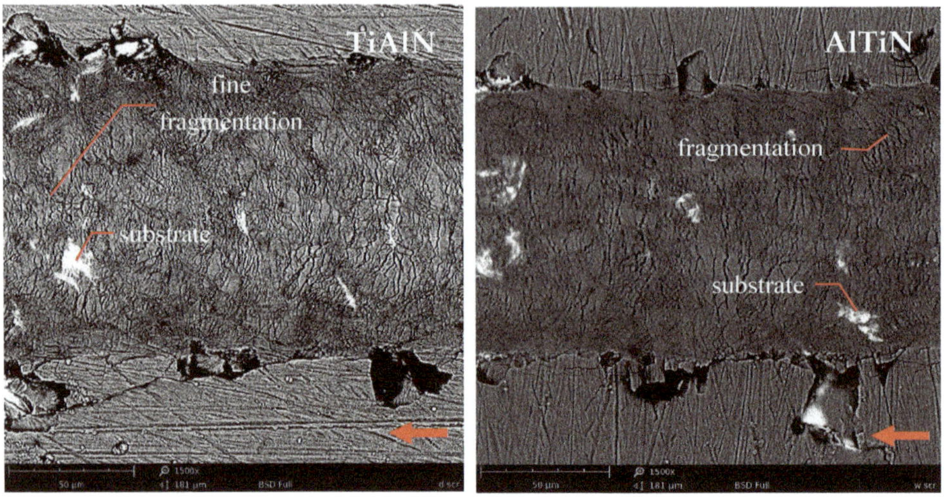

(a) (b)

Figure 4. Comparison of the films at the end of the scratch trace, scratch direction marked by an arrow. (**a**) TiAlN; (**b**) AlTiN, magnif. 1500×, SEM.

In Figure 5, it is clearly visible that the AlTiN and TiAlN coatings have comparable elastic modulus and hardness, in the range of standard deviation. The H and E values are within the result range reported in the literature [9,15,18,24]. Nevertheless, AlTiN exhibits higher H/E and H^3/E^2 ratios than TiAlN, which can be attributed to its higher Al content. Furthermore, the E and H of both films exceed those estimated for the SS substrate.

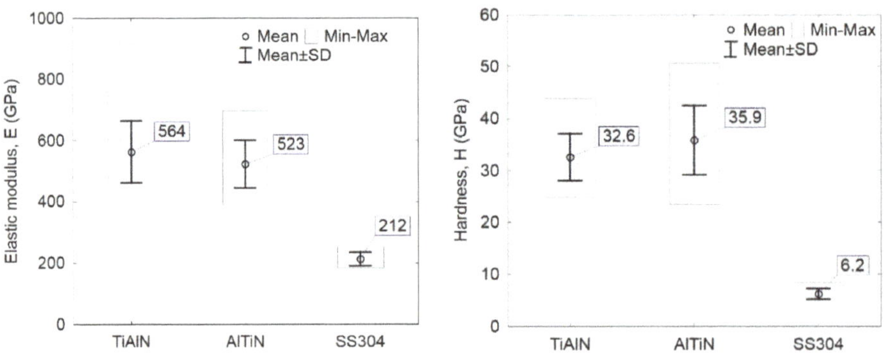

(a) (b)

Figure 5. Elastic modulus and hardness of coated and bare SS304 samples. (a) elastic modulus; (b) hardness.

3.2. Cavitaiton Erosion Resitance (CER)

Quantitative results of the reference stainless steel (SS304) and the magnetron-sputtered AlTiN and TiAlN films deposited on the SS304 substrate are given in Figure 6, whereas a qualitative description of the erosion mechanism is discussed on the basis of Figures 7 and 8 Finally, Figure 9 and Table 3 present the effects of CE on the nanoindentation results, which is an original approach to CER evaluation.

The tested films exhibit higher CERs than the reference SS. The results indicate that the highest CER is achieved by the AlTiN coating, see Figure 6. These results strongly depend on the coating thickness, its mechanical properties and adhesion to the substrate. TiAlN has lower values of thickness, hardness and Lc2. Therefore, it can be deduced that these factors affect quantitative CE results. Also, the H/E and H^3/E^2 coefficients increase and in turn, correlate well with the CER of the TiAlN and AlTiN coatings, which is a new approach to CER evaluation in ternary coating systems. It should be mentioned that a fairly good correlation between the above properties and CER in binary coating systems, i.e., TiN or CrN films, was also shown in the works of Krella [7,33], yet ternary element coatings have not been investigated previously.

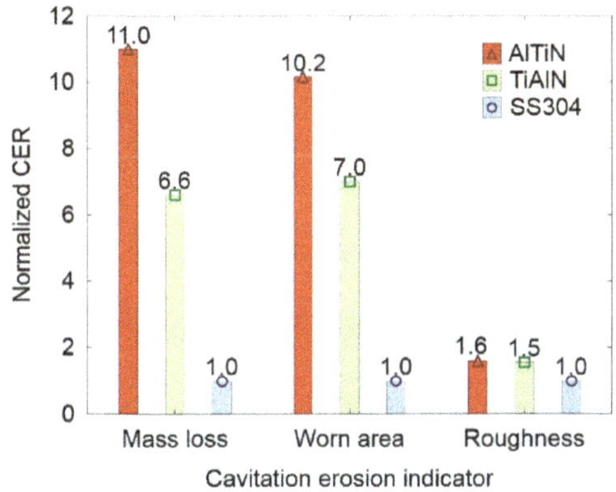

Figure 6. Cavitation erosion resistance (CER) calculated in reference to SS304 based on mass loss (mg), worn area percentage (%) and roughness parameter Sa (μm), 4.5 h of cavitation test.

3.2.1. Mechanism of Cavitation Erosion Damage

A survey of the literature reveals that there are few studies that compare the properties of TiAlN or AlTiN coatings in relation to their CER, and investigate the CE wear mechanism of magnetron-sputtered films with reference to an SS substrate. Figure 7a shows a macroscopic view of surface pitting and Figure 7b compares the surface morphology of TiAlN, AlTiN and SS304 surfaces worn by cavitation erosion. The SEM photographs in Figure 8 provide details into cavitation erosion of the tested materials.

In general, the comparison of the worn surfaces agrees with the quantitative data presented in Figure 6. As a result of cavitation, the sample surfaces are randomly degraded due to the impact of micro-jets or shock pressure waves [28,34,35]. The TiAlN coating is severely more damaged and shows the presence of more pits in its structure than the AlTiN film (Figure 7a). In addition, the observed surface non-uniformities such as "scratches", resulting from substrate preparation, act as surface notches accelerating substrate-perpendicular cracking of the film and its rupture. Nevertheless, both films are less eroded than the reference sample. The roughness parameter *Sa* estimated for the damaged surfaces of TiAlN, AlTiN and SS304 is 1.5 μm, 1.4 μm and 2.3 μm, respectively (Figure 7b).

Measurable film mass losses were obtained after 4.5 h of testing. Thus, the incubation period of CE is approximately three times longer in the case of the hard films than that observed for the bare SS304 samples. This seems promising for increasing the operation time of real stainless steel components coated with magnetron-sputtered TiAlN or AlTiN films.

Generally, the CE mechanism of both AlTiN and AlTiN coatings is brittle by nature and depends on fatigue processes causing coating rupture and flake spallation (Figure 8). However, the deposited TiAlN and AlTiN coatings are much tougher (Figure 5) and brittle than the metal alloy substrate. The fatigue-dependent nature of the CE process causes steel substrate deformation, decreased film adhesion. Films cannot follow substrate deformation. The coating is first spalled and detached in random areas, which is followed by fatigue-induced decohesion of the film, see Figure 8. Once the coating is detached, the density of cavitation pits in the bare surface increases, and the film-deprived areas allow the erosion process to continue into the SS substrate. This CE degradation mechanism observed for TiAlN and AlTiN is in agreement with the findings reported by Krella [7,16] for TiN films (deposited by the cathodic arc technique).

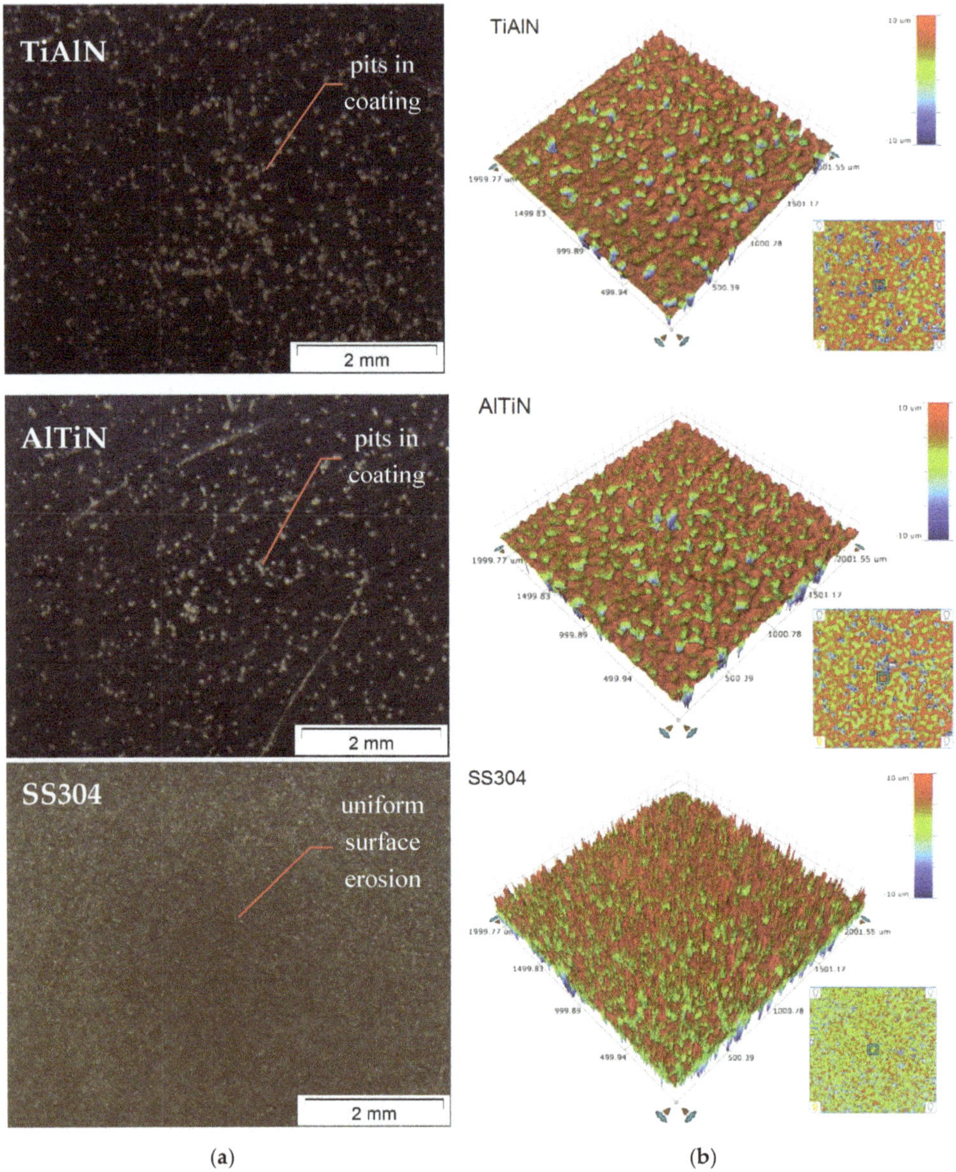

Figure 7. Macroscopic view of cavitation erosion-worn surface, stereoscopic microscope (**a**) and roughness profile of tested surfaces (**b**), after 4.5 h of cavitation testing.

The comparison between the cavitation-worn TiAlN and AlTiN films in pit areas, Figure 8, reveals that the degree of fragmentation is higher in TiAlN than AlTiN. This means that the removal of the fragmented TiAlN coating material is easier, which accelerates its mass loss. In both investigated films, the observed cracks usually propagate along the columnar grain packing, and the dominant CE mechanism is layer flake spallation of the film material. The above are novel findings in the field of CE mechanisms of magnetron-sputtered TiAlN and AlTiN films.

The erosion of the reference stainless steel sample surface is more extensive than that of the coatings, Figure 8. In contrast to the sputtered films, the SS304 surface shows developed damage combining deformation and uniform surface erosion, which is comparable with the results reported in the literature [34–36].

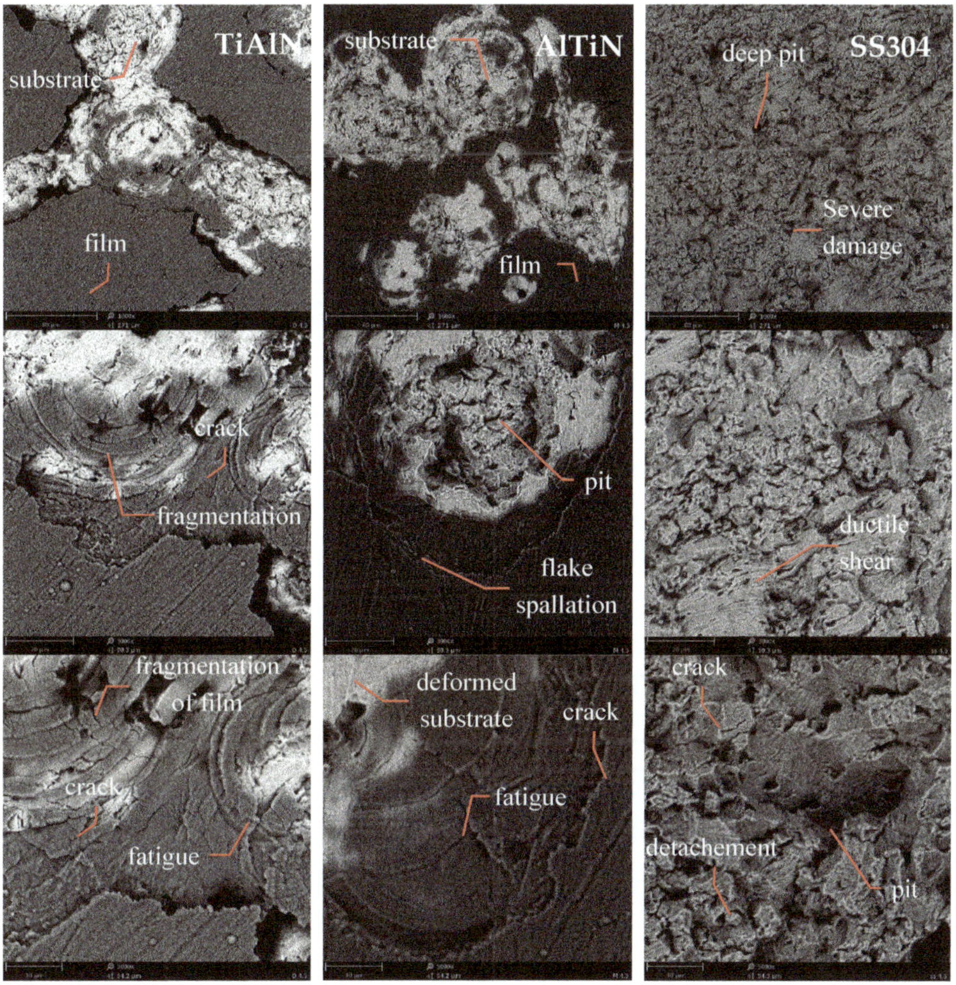

(**a**) (**b**) (**c**)

Figure 8. Comparison of cavitation-eroded thin films and stainless steel, after 4.5 h of testing: column (**a**) TiAlN; (**b**) AlTiN; (**c**) SS304; SEM, 1000×, 3000× and 5000×.

3.2.2. Effect of Cavitation Erosion on Nanoindentation Results

Nanoindentation measurements are a promising method of both CE wear mode identification and CER evaluation. Figure 9 and Table 3 give the results of nanoindentation measurements made before and after cavitation tests, which is a new approach to cavitation erosion testing of thin films. In contrast to the reference SS304 sample, AlTiN and TiAlN show a decrease in the hardness *H* and the elastic modulus *E* after the cavitation erosion tests (Table 3). This finding is in agreement with the results obtained for the SS304 sample, in which one of the first CE effects is usually an increase in surface hardness resulting from work hardening and/or cavitation load-induced phase transformation [34–37]. In the case of the films, however, there occurs fatigue-induced film decohesion occurs followed by the loss of film-substrate adhesion due to cavitation.

A promising method for the ternary films CER identification is to analyze mechanical properties, especially the E/H and H^3/E^2 ratios estimated before cavitation (Table 3). According to the literature [38–42], the susceptibility of a material to elastic strain to failure is expressed by the H/E quotient as well as its resistance to plastic deformation defined by the H^3/E^2 ratio. Higher H/E and H^3/E^2 ratios mean that the films exhibit highly elastic behavior and high resistance to plastic deformation [38–41]. The high value of the H^3/E^2 coefficient is primarily attributed to fracture toughness of thin films. The AlTiN film has a higher H^3/E^2 ratio than the TiAlN film. In other words, the AlTiN coatings are considered to be more resistant than TiAlN, and hence much more resistant to cavitation erosion. This observation is in accordance with the normalized CER results, Figure 6.

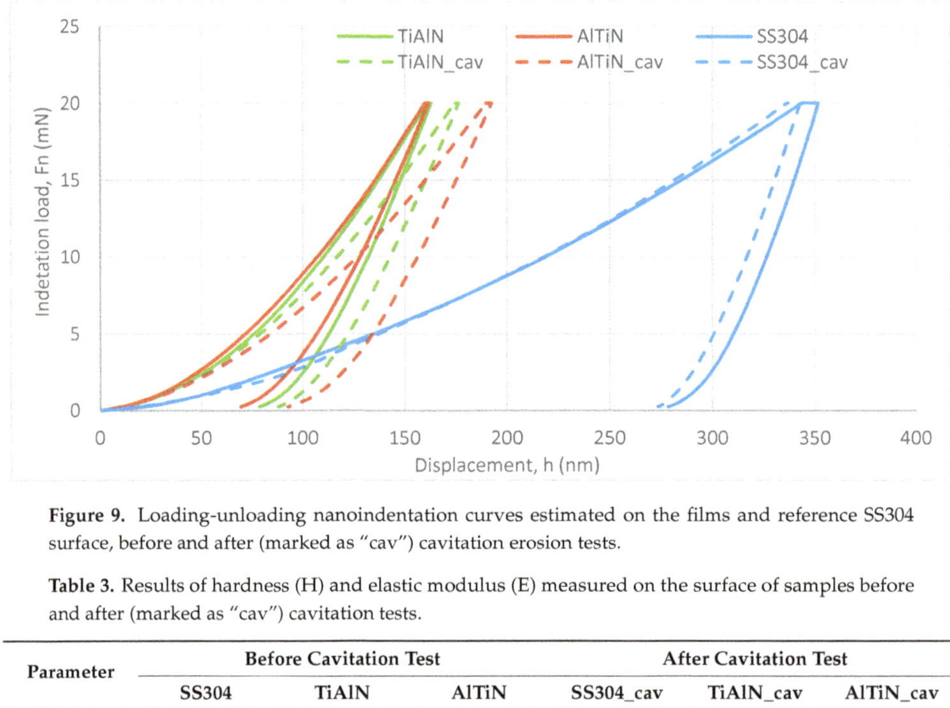

Figure 9. Loading-unloading nanoindentation curves estimated on the films and reference SS304 surface, before and after (marked as "cav") cavitation erosion tests.

Table 3. Results of hardness (H) and elastic modulus (E) measured on the surface of samples before and after (marked as "cav") cavitation tests.

Parameter	Before Cavitation Test			After Cavitation Test		
	SS304	TiAlN	AlTiN	SS304_cav	TiAlN_cav	AlTiN_cav
H [GPa]	6.2 ± 1.0	31.0 ± 4.9	35.8 ± 6.6	6.7 ± 0.3	30.0 ± 6.6	25.4 ± 5.0
E [GPa]	212.4 ± 22.7	564.1 ± 100.9	523.3 ± 77.9	232.8 ± 10.2	452.1 ± 76.9	365.4 ± 57.7
H/E	0.029	0.059	0.068	0.029	0.066	0.070
H^3/E^2	4.012×10^{-6}	6.067×10^{-6}	8.944×10^{-6}	3.558×10^{-6}	9.740×10^{-6}	1.322×10^{-5}

3.3. *Sliding Wear Behaviour (Ball-on-disc Test)*

The ball-on-disc test employed in this study simulates real-life performance under low friction conditions. Quantitative results of the sliding wear test are given in Table 4 and Figure 10. The study has shown that both films present superior sliding wear results (the parameters *K* and μ, see Table 4) to the reference stainless steel sample. In general, in most engineering applications, the higher the H/E and H^3/E^2 coefficients are, the higher the wear resistance of the coating should be [18,42,43]. However, in this study, the differences were also noted between the H/E and H^3/E^2 coefficients of the films, although the resistance to sliding wear of AlTiN and TiAlN is similar (in the SD range, see Table 4). This observation can be explained by low load conditions of the hard films, which affects the relatively shallow wear trace plotted for both coatings, see Figure 10, resulting in a high measuring error of wear.

Furthermore, a qualitative analysis indicates that the SS304 sample undergoes severe sliding wear (see Figures 10a and 11) in comparison to the samples with the deposited TiAlN (Figure 12) and AlTiN (Figure 13). The stainless steel sample shows the presence of adhesive sliding wear and plastic deformation, ploughing with smearing, wear debris oxidation and grooving. Also, due to repeated upsetting and deformation of the material, flake shape delamination with debris transfer resulting from low-cycle fatigue are clearly visible, see Figure 11.

Since the AlTiN and TiAlN films have higher mechanical properties (Figure 5) and resistance to oxidation than the SS sample, their wear mechanism can be classified as grooving and microscratching with the presence of long scratches (Figures 12 and 13). Additionally, microploughing can be observed at the edges of the wear traces of both films. These scratches, groves and ploughs are especially visible in Figure 10b, in the form of non-symmetrical wear profiles. Moreover, the results of debris transfer through the sliding wear trace are visible, and—contrary to the SS sample—no oxidation can be observed in the film wear trace. It can be deduced from the film structures (Figure 2) with the wear trace observed at a higher magnification, see Figure 14, that the sliding wear mechanism consists in smearing the top hills of columnar grains. Generally, after the nano-tribo test conducted under moderate load, the films are pressed toward the SS substrate, and no nitride films delamination can be observed. To sum up, the TiAlN and AlTiN films have similar sliding wear mechanisms.

Table 4. Sliding wear results of the films and reference SS304 sample (mean ± SD).

Sample	Wear Factor, K ($mm^3 \cdot N^{-1} \cdot m^{-1}$)	Coefficient of Friction, μ (–)
TiAlN	$1.35 \times 10^{-5} \pm 4.36 \times 10^{-6}$	0.319 ± 0.037
AlTiN	$2.09 \times 10^{-5} \pm 3.49 \times 10^{-6}$	0.340 ± 0.031
SS304	$50.17 \times 10^{-5} \pm 61.52 \times 10^{-6}$	0.628 ± 0.088

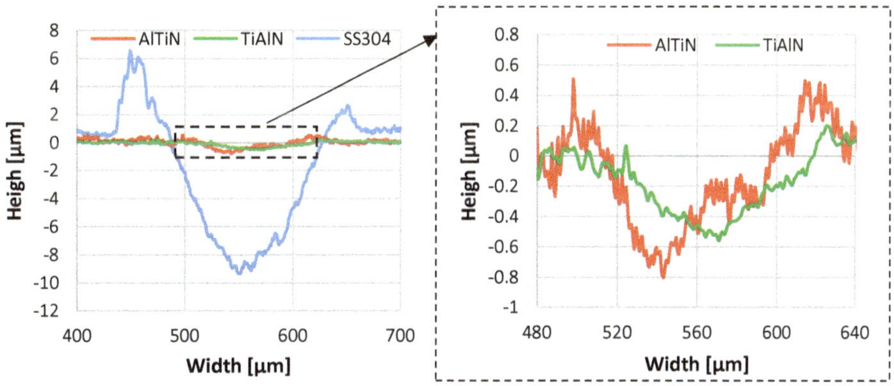

(a) (b)

Figure 10. Sliding wear profiles: (**a**) films and stainless steel sample; (**b**) enlarged selected area of the TiAlN and AlTiN wear traces from (**a**).

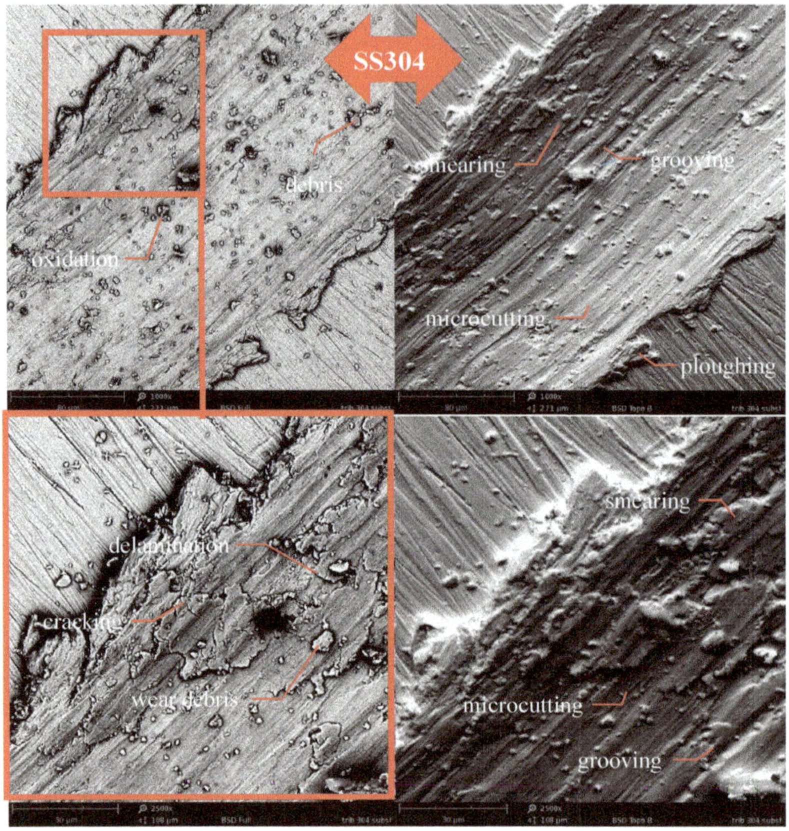

(**a**) (**b**)

Figure 11. Wear trace on the SS304 sample: (**a**) SEM-BSD and (**b**) SEM-topo, 1000× and 2500×.

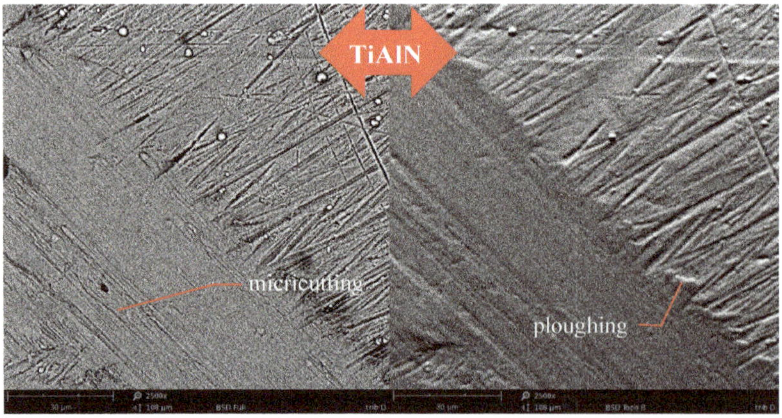

(**a**) (**b**)

Figure 12. Wear trace on the TiAlN film: (**a**) SEM-BSD and (**b**) SEM-topo, 2500×.

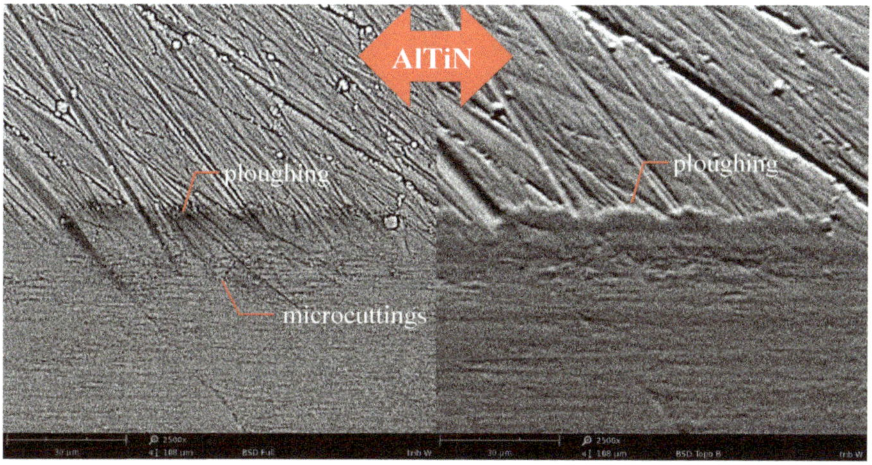

(**a**) (**b**)

Figure 13. Wear trace on the AlTiN film: (**a**) SEM-BSD and (**b**) SEM-topo, 1000× and 2500×.

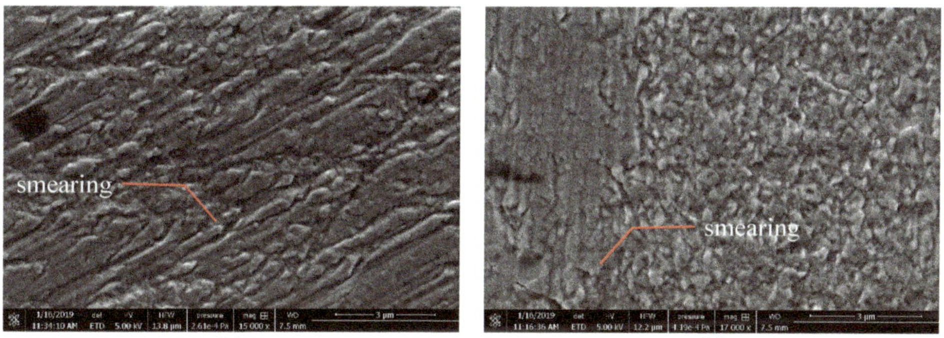

(a) (b)

Figure 14. Sliding wear trace: (**a**) TiAlN; SEM, 15000×. and (**b**) AlTiN; SEM, 17000×.

4. Conclusions

Stainless steel is used for manufacturing different components and considered to be a structural metal with moderate resistance to cavitation erosion. The application of PVD coatings is proposed as a promising solution, easy to implement in industrial practice, for preventing the wear of stainless steel parts. In the present work, the cavitation erosion and sliding wear mechanism of magnetron-sputtered AlTiN and TiAlN coatings deposited on SS304 stainless steel (SS) were investigated. The following conclusions have been drawn:

1. The properties of the films demonstrate that they have a satisfactory structure, i.e., they exhibit a typical columnar morphology comparable to the roughness of the SS substrate, however it varies in thickness, amounting to ≈2.7 μm for TiAlN and ≈3.8 μm for AlTiN. Also, the Rockwell and scratch test results of the films indicate their satisfactory adhesion to the SS substrate, even though a higher force of the Lc2 parameters for AlTiN than TiAlN suggests that AlTiN adheres more strongly to the substrate. The AlTiN film exhibits highly elastic behavior and high resistance to plastic deformation, i.e., its H/E and H^3/E^2 coefficients are higher than those of TiAlN, which can be attributed to its chemical composition, i.e., a higher Al/(Ti + Al) ratio of the AlTiN film.

2. The cavitation erosion resistance of AlTiN is almost one third higher than that of the TiAlN film and almost ten times higher than that of the SS304 sample. The effect of the chemical composition and mechanical properties of the films on their cavitation erosion resistance have been confirmed. The higher Al content, adhesion and H/E and H^3/E^2 coefficients are beneficial for the CER of a film, and can thus serve as a basis for predicting the CER of AlTiN and AlTiN coatings.
3. The cavitation erosion mechanism of both AlTiN and AlTiN coatings is brittle in nature, and depends on fatigue processes that cause rupture and spallation of the coating. However, a comparison of the cavitation-worn TiAlN and AlTiN films reveals a higher degree of fragmentation of the TiAlN film, which ultimately accelerates the wear of the TiAlN film. Additionally, the nanoindentation results (H and E) of the films measured before and after cavitation testing indicate the presence of film material decohesion due to the fatigue-dependent nature of cavitation impact. The film wear mechanism consists in fatigue-induced flake spallation.
4. The sliding wear of the uncoated SS304 sample is higher than that of the sample after PVD coating. The sliding wear resistance of AlTiN and TiAlN is similar yet it is more than 24 times higher than that of the stainless steel sample. Additionally, the application of PVD films onto stainless steel substrates leads to an almost twofold decrease in the friction coefficient. The sliding wear mechanism of the AlTiN and TiAlN films takes the form of grooving, micro-scratching, micro-ploughing and smearing of the columnar grain top hills.
5. The results confirm that cavitation erosion of various fluid machinery components made of austenitic stainless steel can be prevented by depositing AlTiN and TiAlN films.

Author Contributions: Conceptualization, M.S.; Methodology, M.S.; Formal Analysis, M.S. and M.W.; Investigation, M.S., K.P. and M.K.; Writing—Original Draft Preparation, M.S.; Writing—Review & Editing, M.S.; Visualization, M.S., K.P. and M.K.; Supervision, M.S. and M.W.

Funding: The research was financed in the framework of the project Lublin University of Technology—Regional Excellence Initiative, funded by the Polish Ministry of Science and Higher Education (contract No. 030/RID/2018/19).

Conflicts of Interest: The authors declare no conflict of interest.

References

1. Ha, H.-Y.; Jang, J.H.; Lee, T.-H.; Won, C.; Lee, C.-H.; Moon, J.; Lee, C.-G. Investigation of the localized corrosion and passive behavior of type 304 stainless steels with 0.2–1.8 wt % B. *Materials* **2018**, *11*, 2097. [CrossRef]
2. Ławrynowicz, Z. Effect of the degree of cold work and sensitization time on intergranular corrosion behavior in austenitic stainless steel. *Adv. Mater. Sci.* **2019**, *19*, 32–43. [CrossRef]
3. España, P.C.; Recco, A.A.C.; Olaya, J.J. A microstructural and wear resistance study of stainless steel-ag coatings produced through magnetron sputtering. *Coatings* **2018**, *8*, 381.
4. Ha, H.-Y.; Lee, T.-H.; Bae, J.-H.; Chun, D.W. Molybdenum Effects on pitting corrosion resistance of fecrmnmonc austenitic stainless steels. *Metals* **2018**, *8*, 653. [CrossRef]
5. Szala, M.; Beer-Lech, K.; Walczak, M. A study on the corrosion of stainless steel floor drains in an indoor swimming pool. *Eng. Fail. Anal.* **2017**, *77*, 31–38. [CrossRef]
6. Wang, P.; Zhang, Y.; Yu, D. Microstructure and mechanical properties of pressure-quenched ss304 stainless steel. *Materials* **2019**, *12*, 290. [CrossRef]
7. Krella, A.K. The new parameter to assess cavitation erosion resistance of hard PVD coatings. *Eng. Fail. Anal.* **2011**, *18*, 855–867. [CrossRef]
8. Subramanian, B.; Umamaheswari, G.; Jayachandran, M. Properties and corrosion behaviour of reactive magnetron sputtered TiAlN coatings on AISI 316L SS in simulated bodily fluid. *Corros. Eng. Sci. Technol.* **2007**, *42*, 349–355. [CrossRef]
9. Chen, L.; Paulitsch, J.; Du, Y.; Mayrhofer, P.H. Thermal stability and oxidation resistance of Ti–Al–N coatings. *Surf. Coat. Technol.* **2012**, *206*, 2954–2960. [CrossRef]
10. Özkan, D. Friction behavior of TiAlN, AlTiN and AlCrN multilayer coatings at nanoscale. *Erzincan Üniversitesi Fen Bilim. Enstitüsü Derg.* **2018**, *11*, 451–458. [CrossRef]

11. Hans, M.; Patterer, L.; Music, D.; Holzapfel, D.M.; Evertz, S.; Schnabel, V.; Stelzer, B.; Primetzhofer, D.; Völker, B.; Widrig, B.; et al. Stress-dependent elasticity of tialn coatings. *Coatings* **2019**, *9*, 24. [CrossRef]
12. Kulkarni, A.P.; Sargade, V.G. Characterization and performance of AlTiN, AlTiCrN, TiN/TiAlN PVD coated carbide tools while turning ss 304. *Mater. Manuf. Process.* **2015**, *30*, 748–755. [CrossRef]
13. Kohlscheen, J.; Bareiss, C. Effect of hexagonal phase content on wear behaviour of AlTiN Arc PVD coatings. *Coatings* **2018**, *8*, 72. [CrossRef]
14. Fan, Q.-X.; Wang, T.-G.; Liu, Y.-M.; Wu, Z.-H.; Zhang, T.; Li, T.; Yang, Z.-B. Microstructure and corrosion resistance of the AlTiN coating deposited by arc ion plating. *Acta Metall. Sin. Engl. Lett.* **2016**, *29*, 1119–1126. [CrossRef]
15. Shen, W.-J.; Tsai, M.-H.; Yeh, J.-W. Machining performance of sputter-deposited (Al0.34Cr0.22Nb0.11Si0.11Ti0.22)50N50 high-entropy nitride coatings. *Coatings* **2015**, *5*, 312–325. [CrossRef]
16. Krella, A.; Czyżniewski, A. Cavitation erosion resistance of nanocrystalline TiN coating deposited on stainless steel. *Wear* **2008**, *265*, 963–970. [CrossRef]
17. Itoh, T.; Hattori, S.; Lee, K.-Y. Cavitation erosion of 6061 aluminum alloy coated with TiAlN thin film. *J. Solid Mech. Mater. Eng.* **2010**, *4*, 1444–1454. [CrossRef]
18. Łępicka, M.; Grądzka-Dahlke, M.; Pieniak, D.; Pasierbiewicz, K.; Kryńska, K.; Niewczas, A. Tribological performance of titanium nitride coatings: A comparative study on TiN-coated stainless steel and titanium alloy. *Wear* **2019**, *422–423*, 68–80. [CrossRef]
19. Oliver, W.C.; Pharr, G.M. An improved technique for determining hardness and elastic modulus using load and displacement sensing indentation experiments. *J. Mater. Res.* **1992**, *7*, 1564–1583. [CrossRef]
20. Oliver, W.C.; Pharr, G.M. Measurement of hardness and elastic modulus by instrumented indentation: Advances in understanding and refinements to methodology. *J. Mater. Res.* **2004**, *19*, 3–20. [CrossRef]
21. Berg, G.; Friedrich, C.; Broszeit, E.; Berger, C. Development of chromium nitride coatings substituting titanium nitride. *Surf. Coat. Technol.* **1996**, *86–87*, 184–191. [CrossRef]
22. Mattox, D.M. Chapter 10: Film characterization and some basic film properties. In *Handbook of Physical Vapor Deposition (PVD) Processing*; Mattox, D.M., Ed.; William Andrew Publishing: Westwood, NJ, USA, 1998; pp. 569–615, ISBN 978-0-8155-1422-0.
23. Cai, F.; Huang, X.; Yang, Q. Mechanical properties, sliding wear and solid particle erosion behaviors of plasma enhanced magnetron sputtering CrSiCN coating systems. *Wear* **2015**, *324–325*, 27–35. [CrossRef]
24. Walczak, M.; Pasierbiewicz, K.; Szala, M. Adhesion and mechanical properties of TiAlN and AlTiN magnetron sputtered coatings deposited on DMSL titanium alloy substrate. *Acta Phys. Pol. A* **2019**, in press.
25. Vidakis, N.; Antoniadis, A.; Bilalis, N. The VDI 3198 indentation test evaluation of a reliable qualitative control for layered compounds. *J. Mater. Process. Technol.* **2003**, *143–144*, 481–485. [CrossRef]
26. *ASTM G32-10: Standard Test Method for Cavitation Erosion Using Vibratory Apparatus*; West Conshohocken: Philadelphia, PA, USA, 2010.
27. Szala, M.; Hejwowski, T. Cavitation erosion resistance and wear mechanism model of flame-sprayed Al2O3-40%TiO2/NiMoAl cermet coatings. *Coatings* **2018**, *8*, 254. [CrossRef]
28. Szala, M. Application of computer image analysis software for determining incubation period of cavitation erosion—Preliminary results. *EDP Sci.* **2017**, *15*, 06003. [CrossRef]
29. Budzyński, P.; Kamiński, M.; Wiertel, M.; Pyszniak, K.; Droździel, A. Mechanical properties of the stellite 6 cobalt alloy implanted with nitrogen ions. *Acta Phys. Pol. A* **2017**, *132*, 203–205. [CrossRef]
30. Kamiński, M.; Budzyński, P.; Szala, M.; Turek, M. Tribological properties of the Stellite 6 cobalt alloy implanted with manganese ions. *IOP Conf. Ser. Mater. Sci. Eng.* **2018**, *421*, 032012. [CrossRef]
31. Walczak, M.; Pieniak, D.; Niewczas, A.M. Effect of recasting on the useful properties CoCrMoW alloy. *Eksploat. Niezawodn. Maint. Reliab.* **2014**, *16*, 330–336.
32. Kalss, W.; Reiter, A.; Derflinger, V.; Gey, C.; Endrino, J.L. Modern coatings in high performance cutting applications. *Int. J. Refract. Met. Hard Mater.* **2006**, *24*, 399–404. [CrossRef]
33. Krella, A. The influence of TiN coatings properties on cavitation erosion resistance. *Surf. Coat. Technol.* **2009**, *204*, 263–270. [CrossRef]
34. Hattori, S.; Mikami, N. Cavitation erosion resistance of stellite alloy weld overlays. *Wear* **2009**, *267*, 1954–1960. [CrossRef]

35. Mesa, D.H.; Garzón, C.M.; Tschiptschin, A.P. Influence of cold-work on the cavitation erosion resistance and on the damage mechanisms in high-nitrogen austenitic stainless steels. *Wear* **2011**, *271*, 1372–1377. [CrossRef]
36. Wu, Y.; Hong, S.; Zhang, J.; He, Z.; Guo, W.; Wang, Q.; Li, G. Microstructure and cavitation erosion behavior of WC–Co–Cr coating on 1Cr18Ni9Ti stainless steel by HVOF thermal spraying. *Int. J. Refract. Met. Hard Mater.* **2012**, *32*, 21–26. [CrossRef]
37. Szala, M. Coatings for Increasing Cavitation Wear Resistance of Machine Parts and Elements. Ph.D. Thesis, Lublin University of Technology, Lublin, Poland, 2016.
38. Lee, S.-C.; Ho, W.-Y.; Lai, F.D. Effect of substrate surface roughness on the characteristics of CrN hard film. *Mater. Chem. Phys.* **1996**, *43*, 266–273. [CrossRef]
39. Johnson, K.L.; Johnson, K.L. *Contact Mechanics*; Cambridge University Press: Cambridge, UK, 1987; ISBN 978-0-521-34796-9.
40. Tsui, T.Y.; Pharr, G.M.; Oliver, W.C.; Bhatia, C.S.; White, R.L.; Anders, S.; Anders, A.; Brown, I.G. Mechanical behavior of diamond and other forms of carbon. *Mater. Res. Soc. Symp. Proc.* **1995**, *383*, 447. [CrossRef]
41. Musil, J.; Kunc, F.; Zeman, H.; Poláková, H. Relationships between hardness, Young's modulus and elastic recovery in hard nanocomposite coatings. *Surf. Coat. Technol.* **2002**, *154*, 304–313. [CrossRef]
42. Leyland, A.; Matthews, A. On the significance of the H/E ratio in wear control: a nanocomposite coating approach to optimised tribological behaviour. *Wear* **2000**, *246*, 1–11. [CrossRef]
43. Solis, J.; Zhao, H.; Wang, C.; Verduzco, J.A.; Bueno, A.S.; Neville, A. Tribological performance of an H-DLC coating prepared by PECVD. *Appl. Surf. Sci.* **2016**, *383*, 222–232. [CrossRef]

coatings

Article

Stress-Dependent Elasticity of TiAlN Coatings

Marcus Hans [1,*], Lena Patterer [1], Denis Music [1], Damian M. Holzapfel [1], Simon Evertz [1], Volker Schnabel [1], Bastian Stelzer [1], Daniel Primetzhofer [2], Bernhard Völker [1,3], Beno Widrig [4], Anders O. Eriksson [4], Jürgen Ramm [4], Mirjam Arndt [4], Helmut Rudigier [5] and Jochen M. Schneider [1]

[1] Materials Chemistry, RWTH Aachen University, Kopernikusstr. 10, 52074 Aachen, Germany; lena.patterer@rwth-aachen.de (L.P.); music@mch.rwth-aachen.de (D.M.); holzapfel@mch.rwth-aachen.de (D.M.H.); evertz@mch.rwth-aachen.de (S.E.); v.schnabel@web.de (V.S.); stelzer@mch.rwth-aachen.de (B.S.); b.voelker@mpie.de (B.V.); schneider@mch.rwth-aachen.de (J.M.S.)

[2] Department of Physics and Astronomy, Uppsala University, Lägerhyddsvägen 1, 75120 Uppsala, Sweden; daniel.primetzhofer@physics.uu.se

[3] Max-Planck-Institut für Eisenforschung GmbH, Max-Planck-Straße 1, 40237 Düsseldorf, Germany

[4] Oerlikon Surface Solutions AG, Oerlikon Balzers, Iramali 18, 9496 Balzers, Liechtenstein; beno.widrig@oerlikon.com (B.W.); anders.o.eriksson@oerlikon.com (A.O.E.); juergen.ramm@oerlikon.com (J.R.); mirjam.arndt@oerlikon.com (M.A.)

[5] Oerlikon Surface Solutions AG, Oerlikon Balzers, Churer Str. 120, 8808 Pfäffikon, Switzerland; helmut.rudigier@oerlikon.com

* Correspondence: hans@mch.rwth-aachen.de; Tel.: +49-241-80-25980

Received: 4 December 2018; Accepted: 27 December 2018; Published: 2 January 2019

Abstract: We investigate the effect of continuous vs. periodically interrupted plasma exposure during cathodic arc evaporation on the elastic modulus as well as the residual stress state of metastable cubic TiAlN coatings. Nanoindentation reveals that the elastic modulus of TiAlN grown at floating potential with continuous plasma exposure is 7%–11% larger than for coatings grown with periodically interrupted plasma exposure due to substrate rotation. In combination with X-ray stress analysis, it is evident that the elastic modulus is governed by the residual stress state. The experimental dependence of the elastic modulus on the stress state is in excellent agreement with ab initio predictions. The macroparticle surface coverage exhibits a strong angular dependence as both density and size of incorporated macroparticles are significantly lower during continuous plasma exposure. Scanning transmission electron microscopy in combination with energy dispersive X-ray spectroscopy reveals the formation of underdense boundary regions between the matrix and TiN-rich macroparticles. The estimated porosity is on the order of 1% and a porosity-induced elastic modulus reduction of 5%–9% may be expected based on effective medium theory. It appears reasonable to assume that these underdense boundary regions enable stress relaxation causing the experimentally determined reduction in elastic modulus as the population of macroparticles is increased.

Keywords: physical vapor deposition; metastable materials; TiAlN; elastic properties; residual stress; density functional theory

1. Introduction

Metastable cubic transition metal aluminum nitrides such as TiAlN (space group $Fm\bar{3}m$, NaCl prototype) have been well known as hard protective coatings for more than 30 years [1], and are still used nowadays as state-of-the-art materials for cutting and forming applications. However, there is no "the" TiAlN, since materials design enables to tailor the desired properties, such as enhanced thermal stability [2,3]. With respect to a material's bond strength, the elastic modulus connects as design parameter density functional theory-based predictions at the atomic scale with mechanical testing techniques such as nanoindentation. Recently, it has been demonstrated for $Cr_{0.8}Al_{0.2}N$ that a compressive stress state of −4 GPa results in an elastic modulus increase of 150 GPa, when compared to the stress-free material [4]. In addition, compressive residual stresses are utilized to stabilize the metastable cubic phase of MAlN (M = Ti, V, Cr) [5,6].

Protective coatings are usually fabricated by physical vapor deposition (PVD) techniques, such as (high power pulsed) magnetron sputtering or cathodic arc evaporation, and the latter technique is characterized by an almost fully ionized plasma. These techniques are line-of-sight methods, but not limited to materials synthesis on a stationary substrate. Very often, rotating substrates are assembled in industrial batch production plants in order to coat as many substrates as possible in a single deposition process. In such a deposition geometry, the growing coating surface is periodically moved in and out of regions of high plasma density characterized by large fluxes of film-forming species. Recently, we have demonstrated for TiAlON that substrate rotation induces chemical composition modulations, which can be understood by variations in plasma density and fluxes of film-forming species [7]. Going beyond the substrate rotation-induced chemical composition, it appears reasonable that the deposition geometry affects the material performance. Therefore, in the present work we establish the relationship between deposition geometry and mechanical coating properties by means of elasticity, as well as the residual stress state of TiAlN.

2. Materials and Methods

TiAlN coatings were synthesized by cathodic arc evaporation in an industrial-scale Oerlikon Balzers Ingenia P3e™ platform (Balzers, Liechtenstein). Sapphire with (0001) orientation, as well as cemented carbide substrates, were assembled on the substrate holder. Electrically insulating substrates were chosen for comparison to represent ceramic inserts, which can be based on c-BN, Al_2O_3 or Si_3N_4. These ceramic inserts are used for machining of hardened steel, difficult-to-cut materials or cast iron and are frequently coated using PVD techniques in order to reduce tribochemical or adhesive wear.

The deposition system was heated to 450 °C, and the plasma was ignited when the base pressure of the heated system was $<3 \times 10^{-3}$ Pa. Reactive synthesis was carried out with nitrogen gas at the deposition pressure of 3.2 Pa. Three alloyed $Ti_{0.50}Al_{0.50}$ targets were employed, and the minimum target-to-substrate distance was 16 cm. The substrate bias potential was either floating or −15 V. Coatings were synthesized utilizing a two-fold substrate rotation (periodically interrupted plasma exposure) and a stationary deposition geometry without rotation (continuous plasma exposure). The measured substrate temperatures were 470 and 520 °C, respectively, and the stationary deposition setup exhibited an almost factor 10 higher deposition rate compared to the rotation setup.

Additionally, a hybrid setup was employed; substrates were positioned in front of the arc sources and pneumatically controlled shutters were opened and closed periodically according to the effective deposition time in a rotational setup (where the growing coating surface moves periodically from regions of low plasma density/small fluxes of film-forming species in regions with high plasma density/large fluxes of film-forming species). Hence, the hybrid setup is characterized by a face-to-face geometry as in the stationary deposition setup, while variations in plasma density and fluxes of film-forming species (interrupted plasma exposure), as in the rotational setup, are realized by employing shutters.

Furthermore, Fe foil substrates (99.5% purity) were coated with and without substrate rotation in order to obtain powdered coatings. The Fe foil substrates were etched with a nitric acid in a HNO_3/deionized H_2O volume ratio of 1/5 and, subsequently, the coating flakes were milled into a powder. Coatings in powder form were used to determine the linear coefficient of thermal expansion as a function of temperature.

The microstructure of coating cross-sections was characterized in scanning transmission electron microscopy (STEM) mode utilizing a FEI HELIOS Nanolab 660 dual-beam focused ion beam (FIB) microscope (Hillsboro, OR, USA) with a STEM III detector at acceleration voltage and current of 30 kV and 50 pA, respectively. Thin lamellae were extracted in growth direction, synthesized by the rotation, stationary and hybrid setup, utilizing Ga ions with an acceleration voltage of 30 kV. First, a 1-µm-thick Pt protection layer was applied with 80 pA current, followed by trench milling, extraction of the lamella with a manipulation needle and application of the lamella on a Cu Omniprobe. Final lamellae thicknesses in the order of 100 nm were realized by sequential thinning utilizing currents of 0.79 and 0.43 nA.

Depth-resolved chemical composition analysis was done by using time-of-flight energy elastic recoil detection analysis (ToF E-ERDA) with 36 MeV $^{127}I^{8+}$ primary projectiles at the tandem accelerator laboratory of Uppsala University. The measurement geometry of a 45° angle between primary ions and detector telescope and a 22.5° angle between specimen and detector telescope was used. Time-energy coincidence spectra were acquired by combination of a ToF setup based on thin carbon foils [8,9] and utilizing a gas ionization detection system [10]. The obtained spectra were evaluated using the CONTES software package [11]. Homogeneous depth profiles were obtained for Ti, Al and N, while O (<2 at.%) and H (<0.5 at.%) impurities were also detected. Systematic uncertainties were on the order of ±10% (relative deviation), while statistic uncertainties were on the order ±5% (relative deviation). In addition, spatially resolved chemical composition analysis of TiAlN, grown with substrate rotation, was performed utilizing a JEOL JSM-2200FS field emission transmission electron microscope (Tokyo, Japan) at an acceleration voltage of 200 kV. While a high-angle annular dark field (HAADF) detector was employed for morphological characterization, the chemical composition was measured by energy dispersive X-ray spectroscopy (STEM-EDX) with a 30 mm^2 JEOL Si drift detector.

Deposition geometry-dependent roughness values R_a of the coating surfaces were evaluated with a Keyence VK-9700 laser optical microscope (Osaka, Japan). The evolution of roughness data was in good agreement with macroparticle surface coverage values which can be obtained from scanning electron micrographs [12]. Hence, the surface roughness represents an indirect measure of the macroparticle surface coverage.

The resistivity of TiAlN coatings was determined by a Van der Pauw setup [13] using a Keithley 2611B System SourceMeter (Solon, OH, USA) with a current of 5 mA. The measured value for a high power pulsed magnetron sputtered TiN sample was 0.4 µΩ m and in accordance with available literature data in the range of 0.3 [14] to 1.1 µΩ m [15]. All resistivity measurements were done at room temperature and ambient atmosphere.

Elastic properties were investigated by nanoindentation and 100 load-displacement curves were acquired per sample, utilizing a Hysitron TI-900 TriboIndenter (Minneapolis, MN, USA). A Berkovich diamond tip with 100 nm radius was used with a maximum load of 10 mN and the indentation modulus was obtained from the unloading part of load-displacement curves according to the method of Oliver and Pharr [16]. The elastic modulus was calculated from the measured indentation modulus with the Poisson's ratio of $\nu = 0.214$ [17]. Reduction in surface roughness ($R_a < 50$ nm) was realized prior to the nanoindentation tests by mechanical grinding with SiC disks and polishing with diamond suspension.

The phase formation, preferred orientation, and crystallite size were investigated by X-ray diffraction (XRD) using a Siemens D5000 system (Munich, Germany). X-ray source and detector were coupled in θ–2θ scans from 30° to 80° with a step size of 0.04° and a scan time of 4 s per step. The crystallite size was estimated from the (200) peak employing the Scherrer equation [18].

The residual stress state was characterized by XRD stress analysis within a Bruker D8 Discover General Area Diffraction Detection System (Billerica, MA, USA) employing the $\sin^2\Psi$ method and assuming a biaxial stress state [19]. The Cu X-ray source was operated with 40 kV voltage and 40 mA current and the d spacing of the (200) peak was investigated in Bragg-Brentano geometry [20] for tilting angles of Ψ = 0, 18.43°, 26.57°, 33.21° and 39.23° (corresponding to $\sin^2\Psi$ = 0, 0.1, 0.2, 0.3 and 0.4) with respect to the specimen normal. The residual stress was obtained from the slope of the strain

ε plotted as a function of $\sin^2\Psi$ and using elastic modulus values E from nanoindentation experiments according to:

$$\varepsilon = \frac{d - d_0}{d_0} = \sigma\frac{1 + \nu}{E}\sin^2\Psi \quad (1)$$

Interpolation served to determine the stress-free lattice spacing d_0 at Ψ_0 with:

$$\Psi_0 = \sin^{-1}\sqrt{\frac{2\nu}{1 + \nu}} \quad (2)$$

For Poisson's ratio, a constant value of $\nu = 0.214$ was assumed [17]. Stress gradients along the layer thickness as well as deviations from the assumed biaxial stress state affect the accuracy of the calculated stress values.

In situ high-energy XRD measurements on the powdered coatings were performed in transmission geometry at Deutsches Elektronen Synchrotron (DESY) in Hamburg, Germany. Stress-free lattice parameters a_0 were determined from the TiAlN (200) peak and linear coefficients of thermal expansion (CTE) α were calculated according to:

$$\alpha = \frac{1}{a_0}\frac{\partial a_0}{\partial T} \quad (3)$$

Powders grown with substrate rotation and the stationary deposition setup were ground and put into a quartz capillary with a wall thickness of 20 μm and a diameter of 1 mm. The specimens were heated to 600 °C using a Linkam THMS 600 stage (Surrey, UK) in a continuous measurement setup. The stage was constantly purged with Ar to prevent oxidation and powders were illuminated with a monochromatic photon beam (λ = 0.0207 nm). Two-dimensional high energy XRD patterns were acquired every 12 s with a Perkin Elmer 1621 plate detector (Waltham, MA, USA) at a sample-to-detector distance of 771 mm. Data analysis was carried out using the software package FIT2D by integration of the two-dimensional high-energy XRD patterns in q space and conversion to the total structure factor $S(q)$ [21]. Corrections for absorption, fluorescence and inelastic Compton scattering were made.

Density functional theory [22] calculations were carried out utilizing the Vienna ab initio simulation package. Projector augmented wave potentials were used with the general gradient approximation [23] and the ground state was determined by full structural relaxation with convergence criterion of 10^{-3} eV, 500 eV energy cut-off and Blöchl corrections [24]. Total energy minimization was realized as a function of volume with the Birch-Murnaghan equation of states [25] and a $6 \times 6 \times 3$ k-point mesh was used for reciprocal space integration [26]. $2 \times 2 \times 4$ supercells of $Ti_{0.50}Al_{0.50}N$ (128 atoms) were employed and three different configurations were studied: (a) minimum number of Ti-Al bonds (named C#3 in the original work [27]); (b) alternating Ti–N and Al–N layers; and (c) random distribution of Ti and Al.

To obtain the equilibrium volume and bulk modulus as a function of temperature and stress state (configuration with minimum number of Ti-Al bonds), the Debye-Grüneisen model [28,29] was employed, providing the Helmholtz free energy as a function of volume and temperature. By fitting the Helmholtz free energy vs. volume data to the Birch-Murnagham equation of states [25] at each temperature, as well as a volume offset to account for a different stress state, equilibrium volume and bulk modulus were acquired. From these data, elastic modulus, assuming Poisson's ratio of 0.214 [17], and linear coefficient of thermal expansion were extracted.

3. Results and Discussion

The chemical composition of TiAlN coatings with 5 μm thickness, synthesized on sapphire substrates with periodically interrupted (rotation and hybrid deposition geometry) and continuous plasma exposure (stationary deposition geometry), was identified as 27 $\pm$ 2 at.% Ti, 24 $\pm$ 2 at.% Al, 48 $\pm$ 2 at.% N and <1 at.% O, and the geometry-induced differences were in the range of the statistical uncertainty of ERDA. Hence, the deposition geometry/type of plasma exposure does not influence the chemical composition, while the coating microstructure appears to be significantly different, as demonstrated by the STEM micrographs presented in Figure 1. Large columns with widths of up to 500 nm are formed in the case of rotation, see Figure 1a, while finer columns with <100 nm width are observed for the stationary and hybrid deposition geometry, see Figure 1b,c, respectively. Furthermore, scanning electron micrographs of the coating surface (not shown) provide evidence for a strong angular dependence of the macroparticle surface coverage. These findings are consistent with previous reports, see, for example, [30]. The density and size of macroparticles incorporated into the growing coating surface are significantly lower in case of continuous plasma exposure compared to the substrate rotation deposition geometry.

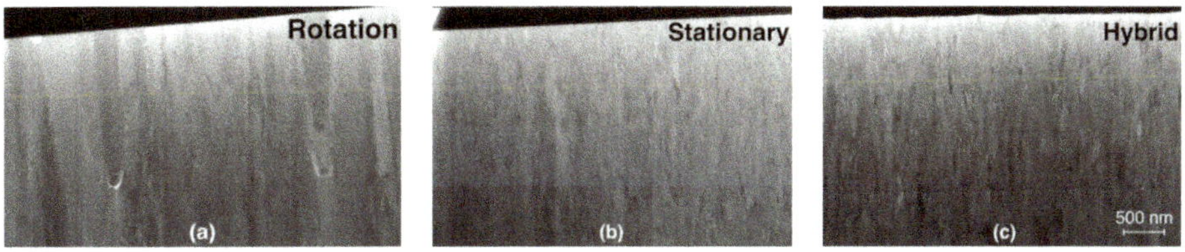

Figure 1. STEM micrographs of TiAlN cross-sections, synthesized with (**a**) substrate rotation, (**b**) stationary, and (**c**) hybrid deposition geometry. All coatings were grown at floating potential and the scale bar in (**c**) is valid for all micrographs.

Diffractograms of TiAlN coatings synthesized with the rotation and stationary setup on cemented carbide substrates (10 μm thickness) are presented in Figure 2, providing evidence for single phase formation of a cubic solid solution structure (NaCl prototype) independent of the deposition geometry. Furthermore, it is evident that the stationary setup results in a stronger (200) preferred orientation compared to the rotation, since the intensity ratios $I_{(200)}/I_{(111)}$ are 4.1 and 13.4, respectively. Crystallite sizes of 35 and 24 nm are estimated for substrate rotation and stationary deposition, respectively. Recently, it has been demonstrated that the extent of cubic and wurtzite TiAlN phase stability regions is predominated by the crystallite size [31]. Based on the present crystallite size data, it is expected that higher critical Al solubilities x in cubic $Ti_{1-x}Al_xN$ can be realized by employing substrate rotation, which results in a larger crystallite size than the stationary deposition geometry.

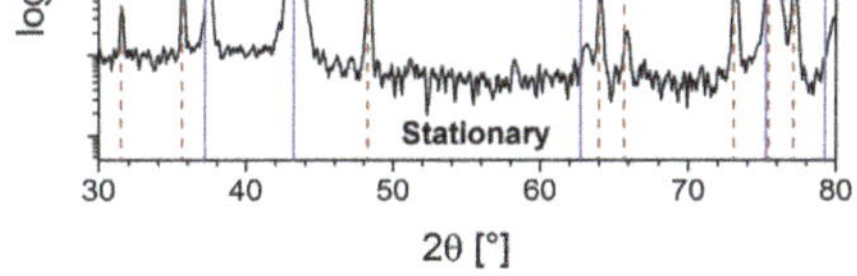

Figure 2. X-ray diffractograms of TiAlN, synthesized with substrate rotation and stationary deposition geometry on cemented carbide substrates (10 μm thickness).

Elastic modulus values of TiAlN are presented as a function of the residual stress state in Figure 3a. Independent of the substrate or coating thickness, 7%–11% higher elastic modulus values are obtained for TiAlN grown with the stationary deposition geometry in comparison to the rotation setup. All samples were grown at floating potential. Depending on the substrate and thickness, residual stress states are in the range of -0.1 ± 0.1 to 1.0 ± 0.1 GPa and -2.3 ± 0.9 to -4.3 ± 1.5 GPa for the rotation and stationary deposition geometry, respectively. The formation of more compressive stress states in case of the stationary setup can be rationalized based on intense fluxes of film-forming species to the growing coating surface due to the almost factor 10 higher deposition rate compared to the rotation setup. It is well known that chemical composition [32], chemical configuration [27], temperature [33], preferred orientation [34] and grain size [35] affect the mechanical properties. However, in the present study, the change in elastic modulus of stationary grown TiAlN as a consequence of composition variations (<1 at.% oxygen incorporation [32]) or chemical configuration (454–457 GPa elastic modulus for the three different configurations) is estimated to be < 1%. Possible effects of growth temperature, preferred orientation and grain size would result in a decrease of elastic modulus when comparing the rotation to the stationary setup, and this is in contrast to the identified increase in elastic modulus. Hence, these factors cannot serve as an explanation for the 7%–11% difference in elasticity obtained by comparing stationary and rotation deposition geometry.

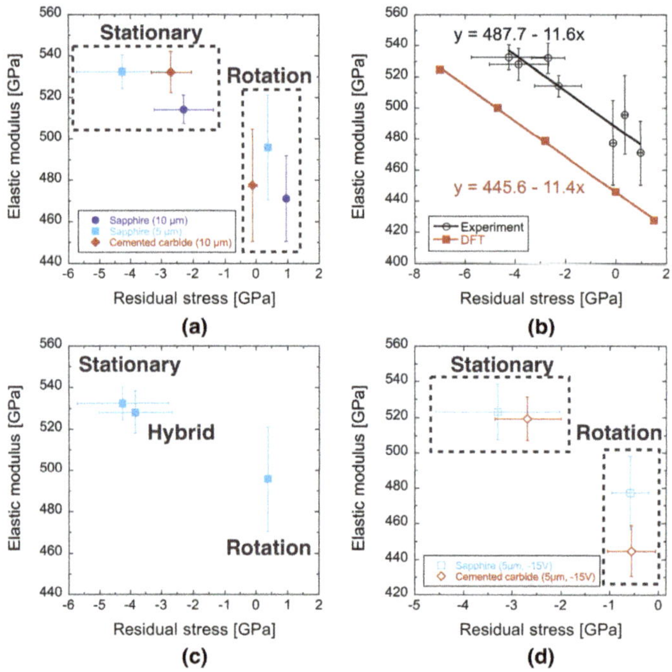

Figure 3. Stress-dependent elasticity of TiAlN, comparing (**a**) the rotation and stationary deposition geometry for different substrate material and film thicknesses, (**b**) DFT predictions (configuration with minimum number of Ti-Al bonds at room temperature) and experimental data, (**c**) the rotation, stationary and hybrid deposition geometry for a thickness of 5 µm on sapphire, and (**d**) the rotation and stationary deposition geometry for different substrate materials at a substrate bias voltage of -15 V. All coatings displayed in (**a**–**c**) were grown at floating potential.

The experimental and theoretical data depicted in Figure 3b clearly illustrate that the elastic modulus is stress-dependent, and that the slope of the predicted data at room temperature is similar to that for the experimental data. The relative deviations of the absolute elastic modulus values are in the order of 9%. This very good agreement between theory and experiment emphasizes that the elastic modulus of metastable cubic TiAlN, grown at floating potential, can be predicted from the residual stress state. The data presented here underline that the deposition geometry affects the residual stress state and, thereby, the elastic properties.

The hybrid deposition geometry can be utilized to compare the interrupted plasma exposure during rotation to the interrupted plasma exposure in the stationary setup. Since both geometries allow for relaxation due to the periodically interrupted plasma exposure, the effect of deposition geometry can be understood. This hybrid setup is characterized by a face-to-face geometry identical to the stationary deposition setup, while shutters enable an interrupted plasma exposure as experienced by the growing film surface during substrate rotation. The stress-dependent elastic modulus data of the stationary and hybrid deposition geometry, presented in Figure 3c, are identical within the error bars. However, compared to the sample grown during rotation, a significant stress-dependent difference in elastic modulus of 7%–8% was obtained. Hence, the constantly high plasma density and intense fluxes of film-forming species during the shutter opening times cause the formation of compressive residual stresses for the hybrid deposition geometry as in the stationary deposition setup, where the plasma exposure is continuous. Therefore, it can be learned that the exposure time of the substrate to the plasma does not influence the residual stress state significantly, since the stress data from the hybrid geometry (with shutter) is within the error bars identical to the stationary geometry (without shutter).

Experimentally, it has been shown that the elastic modulus is stress-dependent for TiAlN and TiAlVN, synthesized by cathodic arc at negative substrate bias voltages ≤ -40 V [36–38]. The associated atomic scale mechanisms have recently been addressed by predictions and experimental verification of $Cr_{0.8}Al_{0.2}N$ taking ion bombardment explicitly into account [4]. To assess the floating potential that is established during continuous vs. periodically interrupted plasma exposure, flat probe measurements were conducted in front of the intense plasma and at the backside of the substrate holder where the substrate is blocked from the plasma due to the holder, see Figure 4a. The substrate position-dependent floating potential is presented in Figure 4b and reveals that the stationary deposition geometry is characterized by a floating potential of -10 to -12 V, while the floating potential during the rotation setup should increase to -4 V, when the substrate is positioned in a region of low plasma density and small fluxes of film-forming species. Therefore, it may be speculated that the constantly "high" floating potential in the stationary deposition geometry explains the 7%–11% higher elastic modulus values compared to the rotation which is characterized not only by variations in the plasma density and fluxes of film-forming species, but also by variations in the floating potential.

This hypothesis was critically appraised by growth experiments utilizing continuous and periodically interrupted plasma exposure with a constant substrate bias potential of -15 V. Provided that the variation in floating potential (during rotation) causes the measured differences in residual stress state and elasticity between coatings produced with rotation and stationary deposition geometry, these differences should be balanced by applying a constant substrate bias potential of -15 V during rotation to mimic the ion bombardment during stationary deposition. Elastic modulus data are presented as a function of residual stress state in Figure 3d for -15 V substrate bias potential and the elasticity of the coatings grown with the stationary deposition geometry is still 10%–17% higher compared to the substrate rotation. Hence, the significant difference in elastic modulus between the two deposition geometries at floating potential cannot be explained by variations in the floating potential during interrupted plasma exposure.

Another well-known contribution to the residual stress state are thermal stresses which are caused by differences in linear coefficient of thermal expansion of film material and substrate at temperatures different from the deposition temperature. The average linear coefficient of thermal expansion (CTE) was calculated from stress-free lattice parameter data, obtained by in situ high-energy XRD. CTE values are with $8.3 \pm 0.4 \times 10^{-6}$ K^{-1} (rotation) and $8.0 \pm 0.4 \times 10^{-6}$ K^{-1} (stationary) in agreement with both published data [39–41] and the predicted value of 7.3×10^{-6} K^{-1}. Therefore, TiAlN exhibits a significantly higher linear CTE than the sapphire substrate of 5.0×10^{-6} K^{-1} (in-plane, provided by substrate manufacturer). When cooling the TiAlN coating from the deposition temperature to room temperature, tensile thermal stresses can be expected, as the coating exhibits larger shrinkage than the substrate. The thermal stress contribution (a comprehensive overview of residual stress contributions is provided in [33]) can be estimated according to:

$$\sigma_{\text{thermal}}(T) = \frac{E_{\text{TiAlN}}}{1 - \nu_{\text{TiAlN}}} \left(\alpha_{\text{sapphire}} - \alpha_{\text{TiAlN}}\right) \left(T - T_{\text{deposition}}\right) \tag{4}$$

Based on this estimation, and taking into account the deposition temperatures of 470 and 520 °C, the thermal residual stress contribution of TiAlN during cooling down from the deposition temperature should be 0.9 and 1.0 GPa for the rotation and stationary deposition geometry, respectively. Interestingly, this estimation is in partial agreement with the measured tensile residual stress values of 0.4 ± 0.1 to 1.0 ± 0.1 GPa in case of the rotation deposition geometry at floating substrate bias potential, but in contrast to the values of stationary grown TiAlN with -4.3 ± 1.5 to -2.7 ± 0.6 GPa, see Figure 3a.

The residual stress state is displayed in Figure 5a as a function of the average surface roughness, which is an indirect measure of the macroparticle surface coverage and thus macroparticle density. It is evident that there are two regions of tensile/slightly compressive residual stress (>−1 GPa) and strongly compressive residual stress (<−1 GPa). Hence, TiAlN coatings grown at floating potential with periodically interrupted plasma exposure during substrate rotation exhibit the largest macroparticle surface coverage and therefore high roughness values. This can be rationalized by considering that the majority of macroparticles are ejected at an angle >10° with respect to the cathode normal [30].

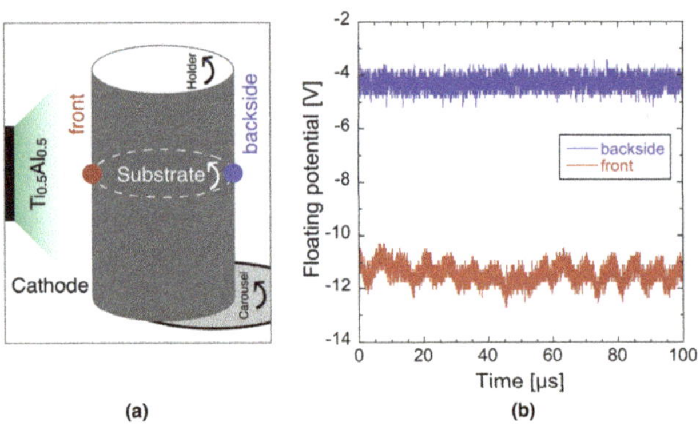

Figure 4. (**a**) Deposition setup scheme indicating the positions for the time-resolved floating potential measurements in (**b**).

To identify the relationship between macroparticle incorporation and the elastic modulus, as well as residual stress state, the cross-sectional morphology and composition of the TiAlN coating with the highest macroparticle surface coverage, highlighted by a circle in Figure 5a, was investigated by TEM and STEM-EDX. From Figure 5b–e, it is evident that underdense (dark) boundary regions are formed between the TiN-rich macroparticles and the TiAlN matrix and an average porosity in the order of 1% was determined from transmission electron micrographs. Based on Knudsen [42], the porosity-dependent elastic modulus can be estimated by:

$$E = E_0 e^{-bP} \tag{5}$$

wherein *E* and *P* represent elastic modulus and porosity, respectively. *b* is an empirical constant and usually assumed to be in the range of b = 5 to 9 [42]. Hence, a porosity-induced elastic modulus reduction of 5%–9% has to be expected. The significance of the macroparticle incorporation-induced porosity is also reflected by the coating resistivity which increases by almost factor 2 from 5.3 to 10.4 μΩ m, comparing TiAlN grown with the stationary and rotation deposition geometry (10 μm thickness), respectively. It appears reasonable that such underdense boundary regions act as sites for stress relaxation. Therefore, it is reasonable to assume that the lower elastic modulus and less compressive/tensile residual stress of TiAlN grown with substrate rotation at floating potential originate from a significantly higher macroparticle surface coverage which induces the formation of underdense boundary regions.

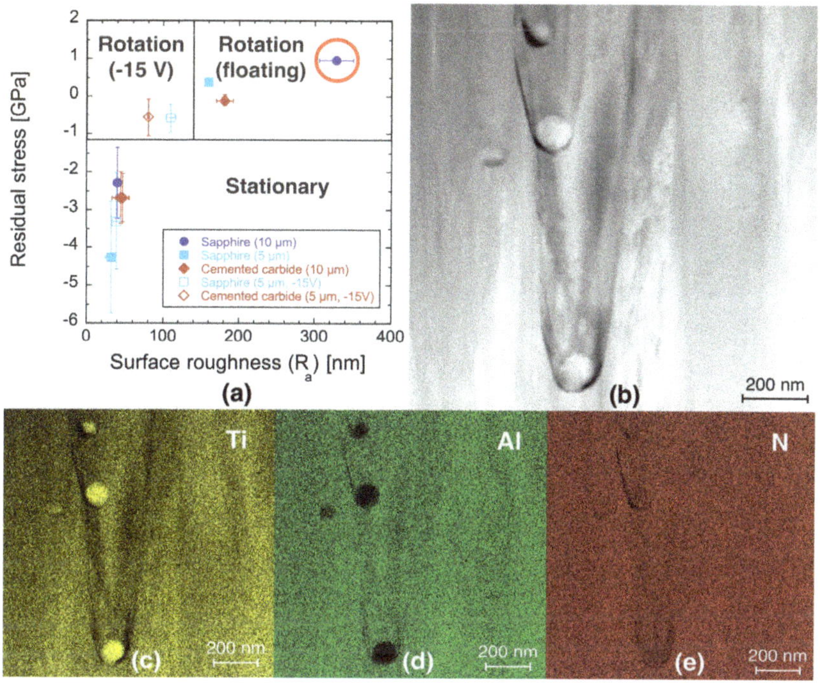

Figure 5. (**a**) Residual stress state as a function of the surface roughness (R_a), which represents an indirect measure of the macroparticle surface coverage. The data point highlighted with a circle indicates the selected coating for (**b**) microstructural and (**c**–**e**) spatially resolved chemical composition characterization (STEM-EDX).

4. Conclusions

We have investigated the effect of continuous vs. periodically interrupted plasma exposure during cathodic arc evaporation on the elastic modulus as well as the residual stress state of metastable cubic TiAlN coatings. The elastic modulus of TiAlN grown at floating potential with continuous plasma exposure is 7%–11% larger than for coatings grown with periodically interrupted plasma exposure due to substrate rotation. Excellent agreement between the experimental and predicted stress-dependent elastic modulus data was obtained. Growth experiments with periodically interrupted plasma exposure due to shutters allowed to infer that the exposure time of the substrate to the plasma is not decisive for the residual stress state. The macroparticle surface coverage exhibits a strong angular dependence, since both density and size of incorporated macroparticles are significantly lower due to continuous plasma exposure. Employing scanning transmission electron microscopy in combination with energy dispersive X-ray spectroscopy, it was revealed that underdense boundary regions are formed between the matrix and TiN-rich macroparticles and the estimated porosity is in the order of 1%. Hence, based on effective medium theory an elastic modulus reduction of 5%–9% has to be expected. It appears reasonable to assume that these underdense boundary regions enable stress relaxation causing the experimentally determined reduction in elastic modulus as the population of macroparticles is increased.

Author Contributions: Conceptualization, M.H. and J.M.S.; Methodology, M.H., D.M., B.S., D.P. and J.R.; Investigation, M.H., L.P., D.M., D.M.H., S.E., V.S., B.S., D.P., B.V., B.W., J.R.; Writing, all the authors; Project Administration, A.O.E., M.A., H.R. and J.M.S.

Funding: This research was partly funded by German Research Foundation (DFG, SFB-TR 87/3) "Pulsed high power plasmas for the synthesis of nanostructured functional layers", Jülich Aachen Research Alliance-High Performance Computing (JARA-HPC, project JARA0151), Swedish Research Council for Research Infrastructures (VR-RFI, contract 821-2012-5144) and Swedish Foundation for Strategic Research (SSF, contract RIF14-0053). Parts of this research were carried out at beamline P02.1 of the light source PETRA III at Deutsches Elektronen-Synchrotron (DESY), a member of the Helmholtz Association (HGF).

Conflicts of Interest: The authors declare no conflict of interest.

References

1. Münz, W.-D. Titanium aluminum nitride films: A new alternative to TiN coatings. *J. Vac. Sci. Technol. A* **1986**, *4*, 2717–2725. [CrossRef]
2. To Baben, M.; Hans, M.; Primetzhofer, D.; Evertz, S.; Ruess, H.; Schneider, J.M. Unprecedented thermal stability of inherently metastable titanium aluminum nitride by point defect engineering. *Mater. Res. Lett.* **2017**, *5*, 158–169. [CrossRef]
3. Schramm, I.C.; Johansson-Joesaar, M.P.; Jensen, J.; Mücklich, F.; Odén, M. Impact of nitrogen vacancies on the high temperature behavior of $(Ti_{1-x}Al_x)N_y$ alloy. *Acta Mater.* **2016**, *119*, 218–228. [CrossRef]
4. Music, D.; Banko, L.; Ruess, H.; Engels, M.; Hecimovic, A.; Grochla, D.; Rogalla, D.; Brögelmann, T.; Ludwig, A.; von Keudell, A.; et al. Correlative plasma-surface model for metastable Cr–Al–N: Frenkel pair formation and influence of the stress state on the elastic properties. *J. Appl. Phys.* **2017**, *121*, 215108. [CrossRef]
5. Holec, D.; Rovere, F.; Mayrhofer, P.H.; Barna, P.B. Pressure-dependent stability of cubic and wurtzite phases within the TiN–AlN and CrN–AlN systems. *Scr. Mater.* **2010**, *62*, 349–352. [CrossRef]
6. Greczynski, G.; Mráz, S.; Ruess, H.; Hans, M.; Lu, J.; Hultman, L.; Schneider, J.M. Extended metastable Al solubility in cubic VAlN by metal-ion bombardment during pulsed magnetron sputtering: Film stress vs. subplantation. *J. Appl. Phys.* **2017**, *122*, 025304. [CrossRef]
7. Hans, M.; to Baben, M.; Chen, Y.-T.; Pradeep, K.G.; Holzapfel, D.M.; Primetzhofer, D.; Kurapov, D.; Ramm, J.; Arndt, M.; Rudigier, H.; et al. Substrate rotation-induced chemical modulation in Ti-Al-O-N coatings synthesized by cathodic arc in an industrial deposition plant. *Surf. Coat. Technol.* **2016**, *305*, 249–253. [CrossRef]
8. Busch, F.; Pfeffer, W.; Kohlmeyer, B.; Schüll, D.; Pühlhoffer, F. A Position-Sensitive Transmission Time Detector. *Nucl. Instrum. Methods* **1980**, *171*, 71–74. [CrossRef]
9. Zhang, Y.; Whitlow, H.J.; Winzell, T.; Bubb, I.F.; Sajavaara, T.; Arstila, K.; Keinonen, J. Detection efficiency of time-of-flight energy elastic recoil detection analysis systems. *Nucl. Instrum. Methods Phys. Res. B* **1999**, *149*, 477–489. [CrossRef]
10. Ström, P.; Petersson, P.; Rubel, M.; Possnert, G. A combined segmented anode gas ionization chamber and time-of-flight detector for heavy ion elastic recoil detection analysis. *Rev. Sci. Instrum.* **2016**, *87*, 103303. [CrossRef]
11. Janson, M.S. *CONTES Instruction Manual*; Uppsala University: Uppsala, Sweden, 2004.
12. Zhirkov, I.; Petruhins, A.; Rosén, J. Effect of cathode composition and nitrogen pressure on macroparticle generation and type of arc discharge in a DC arc source with Ti–Al compound cathodes. *Surf. Coat. Technol.* **2015**, *281*, 20–26. [CrossRef]

13. Van der Pauw, L.J. A method of measuring the resistivity and Hall coefficient on lamellae of arbitrary shape. *Philips Tech. Rev.* **1958**, *20*, 220–224.
14. Ponon, N.K.; Appleby, D.J.R.; Arac, E.; King, P.J.; Ganti, S.; Kwa, K.S.K.; O'Neill, A. Effect of deposition conditions and post deposition anneal on reactively sputtered titanium nitride thin films. *Thin Solid Films* **2015**, *578*, 31–37. [CrossRef]
15. Huang, J.-H.; Yu, K.J.; Sit, P.; Yu, G.-P. Heat treatment of nanocrystalline TiN films deposited by unbalanced magnetron sputtering. *Surf. Coat. Technol.* **2006**, *200*, 4291–4299. [CrossRef]
16. Oliver, W.C.; Pharr, G.M. An improved technique for determining hardness and elastic modulus using load and displacement sensing indentation experiments. *J. Mater. Res.* **1992**, *7*, 1564–1583. [CrossRef]
17. To Baben, M.; Raumann, L.; Music, D.; Schneider, J.M. Origin of the nitrogen over- and understoichiometry in $Ti_{0.5}Al_{0.5}N$ thin films. *J. Phys. Condens. Matter* **2012**, *24*, 155401. [CrossRef]
18. Scherrer, P. Bestimmung der Größe und der inneren Struktur von Kolloidteilchen mittels Röntgenstrahlen. *Nachrichten von der Gesellschaft der Wissenschaften zu Göttingen, Mathematisch-Physikalische Klasse* **1918**, *1918*, 98–100. (In German)
19. Cullity, B.D. *Elements of X-ray Diffraction*, 2nd ed.; Addison-Wesley: Reading, MA, USA, 1956.
20. Brentano, J. Focussing method of crystal powder analysis by X-rays. *Proc. Phys. Soc. Lond.* **1924**, *37*, 184. [CrossRef]
21. Egami, T.; Billinge, S. *Underneath the Bragg Peaks: Structural Analysis of Complex Materials*; Elsevier: Oxford, UK, 2003; Volume 1.
22. Hohenberg, P.; Kohn, W. Inhomogeneous electron gas. *Phys. Rev.* **1964**, *136*, 864–871. [CrossRef]
23. Perdew, J.P.; Burke, K.; Enzerhof, M. Generalized gradient approximation made simple. *Phys. Rev. Lett.* **1996**, *77*, 3865. [CrossRef]
24. Blöchl, P.E.; Jepsen, O.; Andersen, O.K. Improved tetrahedron method for Brillouin-zone integrations. *Phys. Rev. B* **1994**, *49*, 16223. [CrossRef]
25. Birch, F. Finite strain isotherm and velocities for single-crystal and polycrystalline NaCl at high pressures and 300 K. *J. Geophys. Res. Solid Earth* **1978**, *83*, 1257–1268. [CrossRef]
26. Monkhorst, H.J.; Pack, J.D. Special points for Brillouin-zone integrations. *Phys. Rev. B* **1976**, *13*, 5188–5192. [CrossRef]
27. Mayrhofer, P.H.; Music, D.; Schneider, J.M. Influence of the Al distribution on the structure, elastic properties, and phase stability of supersaturated $Ti_{1-x}Al_xN$. *J. Appl. Phys.* **2006**, *100*, 094906. [CrossRef]
28. Söderlind, P.; Nordström, P.; Yongming, L.; Johansson, B. Relativistic effects on the thermal expansion of the actinide elements. *Phys. Rev. B* **1990**, *42*, 4544–4552. [CrossRef]
29. Music, D.; Geyer, R.W.; Keuter, P. Thermomechanical response of thermoelectrics. *Appl. Phys. Lett.* **2016**, *109*, 223903. [CrossRef]
30. Daalder, J.E. Components of cathode erosion in vacuum arcs. *J. Phys. D Appl. Phys.* **1976**, *9*, 2379–2395. [CrossRef]
31. Hans, M.; Music, D.; Chen, Y.-T.; Patterer, L.; Eriksson, A.O.; Kurapov, D.; Ramm, J.; Arndt, M.; Rudigier, H.; Schneider, J.M. Crystallite size-dependent metastable phase formation of TiAlN coatings. *Sci. Rep.* **2017**, *7*, 16096. [CrossRef]
32. Hans, M.; to Baben, M.; Music, D.; Ebenhöch, J.; Primetzhofer, D.; Kurapov, D.; Arndt, M.; Rudigier, H.; Schneider, J.M. Effect of oxygen incorporation on the structure and elasticity of Ti–Al–O–N coatings synthesized by cathodic arc and high power pulsed magnetron sputtering. *J. Appl. Phys.* **2014**, *116*, 093515. [CrossRef]
33. Daniel, R.; Holec, D.; Bartosik, M.; Keckes, J.; Mitterer, C. Size effect of thermal expansion and thermal/intrinsic stresses in nanostructured thin films: Experiment and model. *Acta Mater.* **2011**, *59*, 6631–6645. [CrossRef]
34. Tasnádi, F.; Abrikosov, I.A.; Rogström, L.; Almer, J.; Johansson, M.P.; Odén, M. Significant elastic anisotropy in $Ti_{1-x}Al_xN$ alloys. *Appl. Phys. Lett.* **2010**, *97*, 231902. [CrossRef]
35. Kim, H.S.; Bush, M.B. The effects of grain size and porosity on the elastic modulus of nanocrystalline materials. *Nanostruct. Mater.* **1999**, *11*, 361–367. [CrossRef]
36. Vlasveld, A.C.; Harris, S.G.; Doyle, E.D.; Lewis, D.B.; Münz, W.-D. Characterisation and performance of partially filtered arc TiAlN coatings. *Surf. Coat. Technol.* **2002**, *149*, 217–224. [CrossRef]

37. Ahlgren, M.; Blomqvist, H. Influence of bias variation on residual stress and texture in TiAlN PVD coatings. *Surf. Coat. Technol.* **2005**, *200*, 157–160. [CrossRef]
38. Pfeiler, M.; Kutschej, K.; Penoy, M.; Michotte, C.; Mitterer, C.; Kathrein, M. The influence of bias voltage on structure and mechanical/tribological properties of arc evaporated Ti–Al–V–N coatings. *Surf. Coat. Technol.* **2007**, *202*, 1050–1054. [CrossRef]
39. Tkadletz, M.; Schalk, N.; Daniel, R.; Keckes, J.; Czettl, C.; Mitterer, C. Advanced characterization methods for wear resistant hard coatings: A review on recent progress. *Surf. Coat. Technol.* **2016**, *285*, 31–46. [CrossRef]
40. Bartosik, M.; Holec, D.; Apel, D.; Klaus, M.; Genzel, C.; Keckes, J.; Arndt, M.; Polcik, P.; Koller, C.M.; Mayrhofer, P.H. Thermal expansion of Ti–Al–N and Cr–Al–N coatings. *Scr. Mater.* **2017**, *127*, 182–185. [CrossRef]
41. Tasnádi, F.; Wang, F.; Odén, M.; Abrikosov, I.A. Thermal expansion of quaternary nitride coatings. *J. Phys. Condens. Matter* **2018**, *30*, 135901. [CrossRef]
42. Knudsen, F.P. Dependence of mechanical strength of brittle polycrystalline specimens on porosity and grain size. *J. Am. Ceram. Soc.* **1959**, *42*, 376–387. [CrossRef]

coatings

MDPI

Article

On the Applicability of ATR-FTIR Microscopy to Evaluate the Blending between Neat Bitumen and Bituminous Coating of Reclaimed Asphalt

Alexandros Margaritis [1], Giorgio Tofani [2], Geert Jacobs [1], Johan Blom [1], Serge Tavernier [2], Cedric Vuye [1,*] and Wim Van den bergh [1]

[1] EMIB Research Group, Faculty of Applied Engineering, University of Antwerp, Groenenborgerlaan 171, 2020 Antwerp, Belgium; alexandros.margaritis@uantwerpen.be (A.M.); geert.jacobs@uantwerpen.be (G.J.); johan.blom@uantwerpen.be (J.B.); wim.vandenbergh@uantwerpen.be (W.V.d.b.)

[2] BioGEM Research Group, Faculty of Applied Engineering, University of Antwerp, Salesianenlaan 90, 2060 Antwerp, Belgium; giorgio.tofani@uantwerpen.be (G.T.); serge.tavernier@uantwerpen.be (S.T.)

* Correspondence: cedric.vuye@uantwerpen.be; Tel.: +32-495-440707

Received: 21 March 2019; Accepted: 6 April 2019; Published: 9 April 2019

Abstract: The utilization of Reclaimed Asphalt (RA) in the road construction sector induces considerable economic and ecological benefits. The blending of the recycled material with new components is believed to be of great importance for the mixture's properties. An extensive knowledge of the blending of the materials is crucial in optimizing the use of RA, especially at higher recycling rates. In this paper, the applicability of Fourier transform infrared (FTIR) microscopy in attenuated total reflectance (ATR) mode to study the bituminous coating of RA granulates is investigated. This method is a promising alternative to trace heterogeneous areas within the coating compared to methods that require extraction and recovery of bitumen. A method for sample preparation and FTIR spectra analysis is proposed. Four different samples were analyzed: a reference RA granulate, two types of RA granulates mixed with neat bitumen, and a RA granulate with rejuvenator. The results show that the use of ATR-FTIR microscope allows the tracing of different components, indications of blending, as well as proof of rejuvenation of the aged bituminous area.

Keywords: reclaimed asphalt; asphalt recycling; ATR-FTIR microscopy; chemical imaging; ageing; bituminous blending; rejuvenator

1. Introduction

When asphalt pavements reach their end-of-life point, their layers must be renewed. The material removed from these layers is known as Reclaimed Asphalt (RA) (EN 13108-8:2016, [1]). RA is in fact an asphalt mixture consisting of aggregates and bitumen. The removed material is typically crushed in agglomerations of different sizes. Asphalt recycling is not only economically viable, since a considerable amount of bitumen can be replaced by the RA bitumen in the new asphalt mixture, but also environmentally friendly [2]. Previous research has shown the potential benefits of using RA on the mechanical performance of bituminous mixtures [3] and also provided recommendations concerning the proper exploitation of the material [4].

When a new bituminous mixture with the incorporation of RA is designed, in most design procedures, the assumption of full blending between the aged RA bitumen and the new bitumen is still being used. The mobilization or activation of RA bitumen is, till today, a "black box" for the asphalt sector; the actual degree of blending between new and aged bitumen is under research. Many studies question this practice and have demonstrated cases of partial blending [5,6] or zones where the phenomenon of "black rock" is present, meaning that part of the bitumen in the RA is inactive [7].

This gap in the scientific knowledge is of great importance, since overestimating the degree of blending can lead to mixtures with less active bitumen, which could significantly influence the mechanical performance of bituminous mixtures [8].

Previous studies have approached this problem in different ways. The first approach is by investigating the diffusion between two bituminous layers and studying the response of the blend by means of Dynamic Shear Rheometer (DSR) [9,10]. Other studies have modelled the activated bitumen modulus response by using the stiffness modulus of the corresponding mixture [11]. Researchers have also performed staged extractions to RA granulates in order to assess the properties of the recovered bitumen layer by layer in terms of shear modulus and infrared spectra analysis [12–14]. The latter approach contains some risks for interpretation of the results, as it is still unclear whether the solvents influence the bitumen structure and what their impact is on the properties of the recovered binder [15].

The past years, increasing attempts have demonstrated the ability of non-destructive tests to evaluate this problem. Attenuated total reflectance (ATR) spectra obtained by Fourier transform infrared (FTIR) spectroscopy is a widely utilized non-destructive test to study bitumen and especially to trace ageing evolution [16–18], to evaluate rejuvenation [19,20] or to evaluate the modification of bitumen [21,22]. ATR-FTIR spectroscopy is suitable to evaluate homogeneous individual components but not a heterogeneous composite material (such as an asphalt mixture), since it can only be applied in a single spot averaging the measuring spectra. For that purpose, more and more researchers draw attention to FTIR microscopy techniques.

ATR-FTIR imaging has been used on asphalt concrete samples with RA in order to demonstrate a method to evaluate the different components highlighting that sample imperfections strongly effect the quality of the measurements [23]. Besides ATR-FTIR imaging, X-ray fluorescence and infrared microscopy have been used in an additional study. The analysis was applied on the bitumen level using markers to track the mobilization of artificially aged bitumen. The conclusion was that ATR imaging was the most successful technique [24]. Other studies have evaluated the blending of asphalt concrete enriched with titanium dioxide (TiO_2) as a tracer using Computer Tomography (CT), in macro- and micro-scale [25], and using environmental scanning electron microscopy (ESEM) [7,25]. A scanning electron microscope/energy dispersive spectrometer (SEM/EDS) evaluated the degree of blending of mixtures, with various RA quantities, again using TiO_2 as a tracer [26]. Research has also been conducted using atomic force microscopy (AFM) on bitumen levels [26,27]. Another attempt at investigating the blending between two different bituminous binders was by nano-CT scanning images and nano-indentation tests, which did not provide clear and direct conclusions [28].

In order to study the blending process between aged and new bituminous layers, previous research simulated the ageing state of the RA bitumen with artificially aged bitumen in the laboratory; it has been demonstrated that long-term laboratory ageing techniques do not always correspond to field ageing [29,30]. On the other hand, several studies have used actual RA but with the addition of tracers to mark the mobilization, such as the TiO_2. The question that might arise here is to what extent markers can influence the mobilization. Furthermore, earlier studies considered the mobilization of only bitumen or focused only on the diffusion process between two bituminous binders. The aforementioned procedures highlight the need of evaluating field-aged asphalt materials and focus on the actual blending conditions as it will happen in the bituminous coating of an RA granulate. In reality, the stones in the asphalt mixture are bonded not by pure bitumen but by the bituminous mortar (Figure 1). For clarification purposes, the following definitions are given: asphalt mortar is a combination of bitumen and fine particles (filler and sand smaller than 0.5 mm) and asphalt mastic is a combination of bitumen and filler.

In this study, actual RA granulates are used without the addition of foreign tracers and provide the opportunity to study the spatial distribution of neat bitumen within the mortar scale by means of ATR-FTIR microscopic measurements. This research aims firstly to assess the application of FTIR microscopy directly on actual RA granulates, secondly to fingerprint the different zones of components, and thirdly to trace rejuvenating agents and mobilization of neat bitumen within the coating of the granulates.

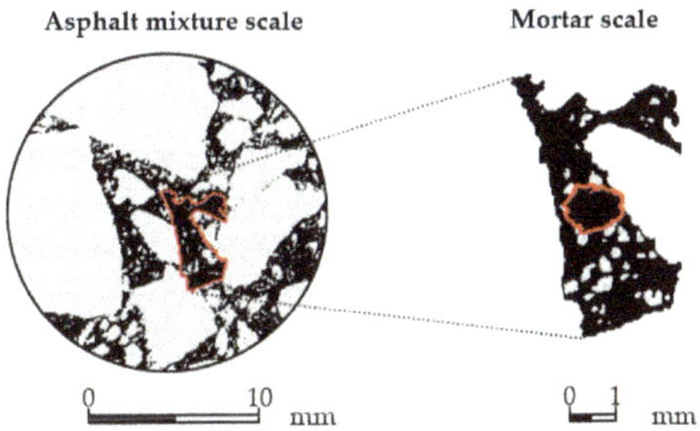

Figure 1. Asphalt mixture bonded by asphalt mortar.

2. Materials and Methods

For the purpose of achieving the aims of this research, an FTIR microscope (LUMOS by Bruker, Billerica, MA, USA) in ATR-mode was used to evaluate the coating of various RA samples. Characteristic absorbance bands were selected, which were typical for the components that form the final sample, i.e., bitumen, fines, rejuvenator, and resin (see Section 2.3). After the definition of the spectra, which is a graphical visualization of the infrared light absorbance as a function of the wavelength in the mid-infrared region of 4000—600 cm^{-1}, data concerning the integration of the characteristic areas were analyzed and formed into a chemical imaging of the area studied (see Section 2.3). The integrated areas were further analyzed to inspect the numerical distribution of carbonyl levels within the bituminous zone (Section 3.2).

2.1. Sample Preparation

Four types of RA granulate samples were studied in this study. Among the samples, one is an untreated RA granulate (sample name: GRA) and three are treated RA granulates (sample names: GRAB, MRAB, and GRAR). A summary of the samples and their treatment is presented in Table 1. The RA used in this study consists of granite and limestone aggregates (inspected visually), and the bitumen after extraction and recovery has a penetration of 24 1/10 mm. The treated RA granulates were fabricated following three different procedures. First, sample GRAB was prepared by mixing the RA granulates with neat bitumen. Second, sample MRAB is an RA granulate that was derived from an actual asphalt concrete mixture (type AC14, see Table 2). Third, the GRAR sample was sprayed with rejuvenator. The preparation of the treated samples is more elaborately described in the following paragraphs.

Table 1. Samples and treatment description.

Sample Name	RA Granulate	Neat Bitumen	Rejuvenator	Treatment Description
GRA	✔	–	–	Untreated RA granulates (fraction 10/12 mm)
GRAB	✔	✔	–	RA granulates covered with neat bitumen (fraction 10/12 mm)
MRAB	✔	✔	–	RA granulates, covered with neat bitumen, obtained from mixed AC-14 with GRA
GRAR	✔	–	✔	Rejuvenated RA granulates

Table 2. Composition of the AC14 mixture.

Component	Percentage
RA aggregates 0/14	55.7%
Limestone 4/8	18.4%
Limestone 2/6.3	6.4%
Limestone 0/2	18.5%
Limestone filler	1.0%
RA bitumen	3.0%
Neat bitumen (168 1/10 mm)	1.3%

The GRAB samples were fabricated following the procedure described in § 5.2 of EN 12697-11:2012 [31]. In this standard, a method is described to create a uniform bituminous coating on mineral aggregates. An adequate amount of RA granulates (600 g) was preheated at 110 °C. In order to avoid excessive ageing of the RA binder, the preheating time was limited to two hours. The preheated granulates were then manually mixed together with 3% of 35/50 penetration grade neat bitumen (by RA granulates mass). The mixing conditions were a maximum of 2 min at 150 °C. Afterwards, the covered stones were spread and separated on silicone paper (Figure 2b). Visually traced agglomerations were rejected.

Figure 2. Original, untreated Reclaimed Asphalt (RA) granulates (**a**); RA granulates mixed with neat bitumen (**b**); RA granulates sprayed with rejuvenator (**c**).

The MRAB samples were obtained from an actually mixed AC14 mix with 55.7% of RA in the mixture. The composition of the mixture was designed in such way that the only source of coarser aggregates (larger than 8 mm) was the RA granulates. The mix composition is presented in Table 2. The production of the mixture was done according to EN 12697-35:2016 [32]. In contrast to the GRAB samples, the mixing took place for 5 min at a temperature of 180 °C instead of maximum 2 min at 150 °C. After mixing, only the larger stones (>8 mm) were selected for MRAB in order to assure only covered GRA was sampled.

Finally, the GRAR samples were prepared by spraying a crude tall oil-based rejuvenator on the samples. Six (6) RA granulates were sprayed with 2% rejuvenator on the RA mass (Figure 2c). Both RA and rejuvenator were at room temperature. In order to accelerate the diffusion of rejuvenator [33], the sprayed stones were placed for 45 min in an oven at 110 °C. This temperature is according to the heating limits of RA proposed in § 6.3 of EN 12697-35:2016 [32]. The amount of rejuvenator added can be considered as rather high when compared to other studies [34]. The purpose here is to make sure there is a traceable amount of rejuvenator.

A necessary step for the utilization of the FTIR microscope is the existence of a flat and even sample surface. In order to fulfil this requirement, the RA granulates were embedded in epoxy resin. For the preparation of the stone-resin samples, the following steps were followed: First, the samples were embedded in a plastic mold with resin (Figure 3a). The height of the stone-resin sample is

25 mm and the diameter 30 mm. The samples were then cured at room temperature for one day. Afterwards, the cured samples were demolded and polished with a mechanical grinding/polishing machine (Struers TegraForce 5) (Figure 3b). The polishing was done in two steps. During the first step, a 220 grit size sandpaper removed the larger part. During this step, the plateau is rinsed by cool water, which keeps the temperature low during the polishing. The second step uses a high friction paper together with a diamond paste (9 μm), providing an even final surface.

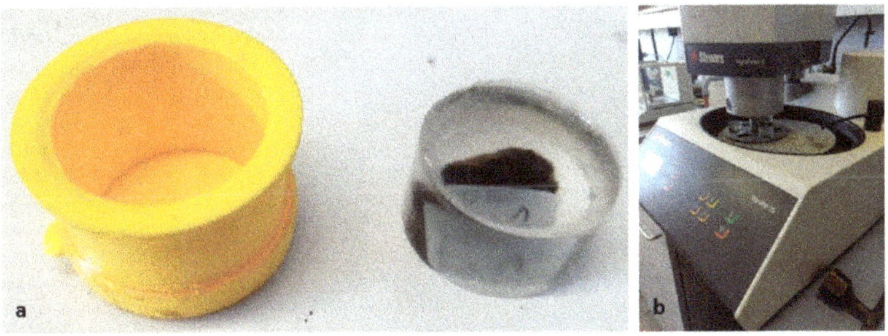

Figure 3. Plastic mold for preparation of the stone-resin samples (**a**) and the polishing machine (**b**).

2.2. FTIR Instrumentation

In this study, an FTIR Microscope (Lumos by Bruker, Billerica, MA, USA) was used. The infrared analyses were performed using a germanium ATR crystal with a high refractive index ($n_{Ge} = 4$), which permits the analysis of dark samples. The instrument is able to generate a chemical map by integrating the infrared spectra over a specified area. The integration methods used for this study are presented in Section 2.3. The stone-resin samples were placed on a mounting holder called the Micro-Vice Holder (Figure 4). Each spectrum was compiled from 32 scans with a resolution of 4.0 cm^{-1} in a range of 4000–600 cm^{-1}. To avoid damaging the crystal, the applied contact pressure was limited to "low pressure".

Figure 4. Stone-resin sample mounted in the holder during spectra collection.

2.3. Methodology

For the investigation of the RA granulates' coating, a procedure was set (Figure 5). First, the area to be measured is selected. Second, the grid of the points to be measured is defined. Third, the measured spectra are preprocessed. Fourth, the integrated typical bands are visualized. The intensity visualization of the bands on the microscopic image is defined as chemical imaging. The processing of the spectra and evaluation of the derived data was done using the Bruker OPUS™ software (v. 7.5). Each sample was analyzed following the proposed steps.

Figure 6 shows the polished surface of the samples and the areas selected to be analyzed (Areas 1 to 5). This figure shows only the exposed surface of the granulate without the resin. For illustrative purposes, the resin was removed graphically in Figure 6 to provide a clear view of the samples and their bituminous coating. The areas were selected with the purpose of observing potential traces of partially or non-blended aged bitumen and to track the presence of other components such as resin or fine particles. The focus was mainly on the interface of mortar and resin (Areas 2 and 4), and also on areas starting from the resin towards the stone, creating a profile (Areas 3 and 5). Area 1 is a single case where the analysis was performed in a significantly larger area. On average, the spatial resolution applied for each point of the analysis was 30×30 μm^2.

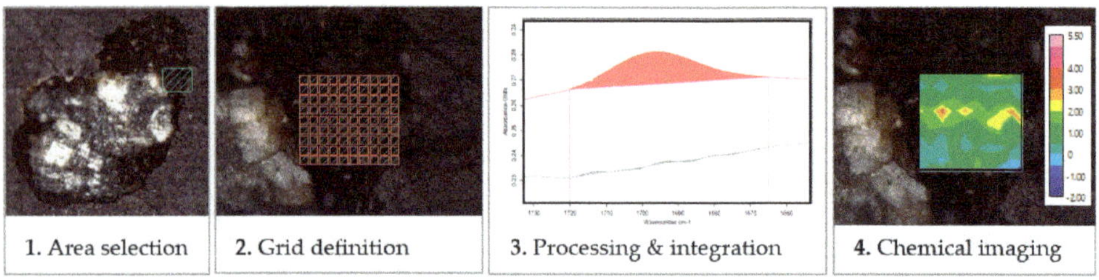

Figure 5. Procedure for determination of the chemical imaging.

Figure 6. Samples and selected areas for analysis: Area 1 in GRA sample, Areas 2 and 3 in GRAB sample, Area 4 in MRAB sample, and Area 5 in GRAR sample.

The solidity of the studied samples does not allow probe penetration into the sample, which has a direct effect on the intensity of the absorbance signal. The software provides the solution of atmospheric compensation, which allows the correction of the signal. This step uses "physical models to estimate the amount of atmospheric gases in the single-channel spectra and therefore compensates disturbing H_2O and/or CO_2" [35]. The compensated areas are from 3600 to 4000 cm^{-1} for H_2O compensation and from 2300 to 2400 cm^{-1} for CO_2 compensation [35]. This step only affects the aforementioned bands, and it is suitable for the current research since those areas are free of absorption bands related to the RA samples (see Table 3).

The spectra were normalized using the min-max normalization method [35]. This method was applied by selecting the C–H stretching band (2990–2820 cm^{-1}) as the band to be maximized. The result is that the absorbance of the most intense peak (at 2920 cm^{-1}) is scaled to 2. A normalization step is required in order to avoid intensity problems caused by varying contact between the sample and the crystal. The C–H stretching band is the most appropriate since it is not affected by bitumen ageing [16] and it is present for both bituminous and resin zones.

For the visualization of the chemical images, specific bands were selected to be integrated. The integration method was used to quantitatively measure the area under the spectrum graph, which represents the intensity of a certain functional group, using wavelength limits as a baseline. The characteristic bands used in this study for each component are presented in Table 3. The tracing of the carbonyl group was the main focus since the presence of carbonyls indicates ageing of bitumen [17]. The carbonate group was used as a marker of the mastic/mortar area and the carboxylic acids group was used for the rejuvenator. Finally, the resin can be distinguished by the ring vibration of the epoxides bond. Figure 7 demonstrates the integration method of the characteristic infrared bands, integrated in spectra derived from different materials by means of FTIR-ATR spectroscopy. The chemical images can then be formed on the sample microscopic image, providing the opportunity to trace spatial intensity of the integrated functional groups.

Table 3. Distinctive bands of the individual components.

Bond & Functional Group	Peak Point (cm^{-1})	Baseline Limits (cm^{-1})	Presence in Material
C–H stretch (alkyl bonds)	2920, 2848	2990–2820	Bitumen & Resin
C=O stretch area (carboxylic acids)	1741	1760–1720	Rejuvenator
C=O stretch area (carbonyl)	1700	1720–1665	Aged bitumen
C–O vibration (carbonate)	875	890–850	Limestone
C–O–C (ring vibration of epoxides)	829	858–785	Resin

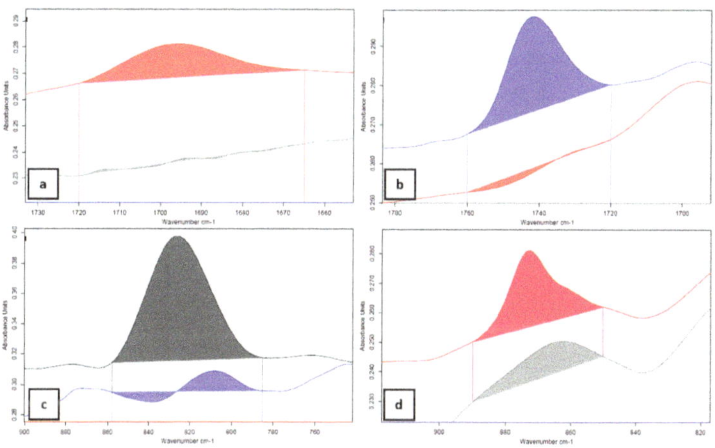

Figure 7. Carbonyl of RA bitumen (red) and neat bitumen (light grey) (**a**); carboxylic acids of rejuvenated bitumen (blue) and non-rejuvenated bitumen (red) (**b**); epoxides of resin (dark grey) and of bitumen (blue) (**c**); carbonate of bitumen with limestone filler (magenta) and bitumen without filler (light grey) (**d**).

3. Results and Discussion

Five surface areas of RA granulates were analyzed using FTIR microscopy in ATR mode. Using the infrared spectra derived, chemical images were formed by integrating the typical bands for each area separately. In Section 3.1 the spatial distribution of typical bands is presented, and in Section 3.2, the numerical distribution of carbonyl levels is exhibited only for the bituminous area.

3.1. Chemical Imaging

The selected areas (Areas 1 to 5) cover three basic situations: (i) original ageing state of RA bitumen, (ii) interaction between neat bitumen and RA, and (iii) rejuvenation of RA. In order to have comparable chemical imaging, the color plot scaling is the same between chemical maps of the same band, except for the carbonate group, since larger variations in absorbance are noticed here compared to other areas.

Figure 8 shows the results of the chemical imaging analysis of the GRA sample. The spatial resolution of the area studied is 9800 × 5400 μm^2. In Figure 8a, the integration of carbonyls is demonstrated, while Figure 8b,c show the carbonate and the epoxides, respectively. Here, it is noticeable that the resin area is distinguishable by the chemical imaging of the epoxide (level above 60). The level of carbonyl varies here between 0 and 5.5. In Figure 8b, larger aggregate particles (white areas) and also mortar areas can be seen from the integration of the carbonate.

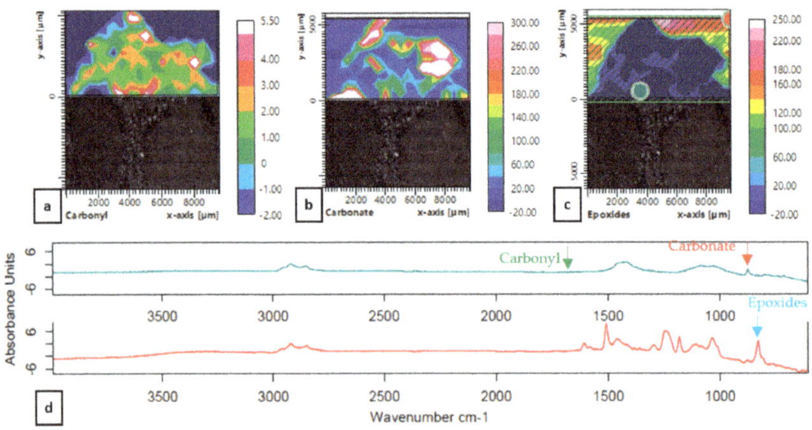

Figure 8. Chemical imaging of the GRA sample (Area 1): carbonyl group (**a**); carbonate group (**b**); epoxides (**c**); spectral visualization of the selected points marked by colored dots in Figure 8c (**d**).

The chemical imaging analysis of Area 2 of the GRAB sample is presented in Figure 9. The spatial resolution of Area 2 is 390 × 2450 μm^2. Figure 9a shows that the carbonyl level is lower compared to sample GRA (Figure 8a) and that the main part of Area 2 is resin free, except from a minor area at the upper side (Figure 9c). According to Figure 7a, the absence of carbonyls is an indication of unaged bituminous areas. Based on that observation and by detecting the carbonyl level of Area 2, the presence of a blend between aged and neat bitumen can be assumed in this area.

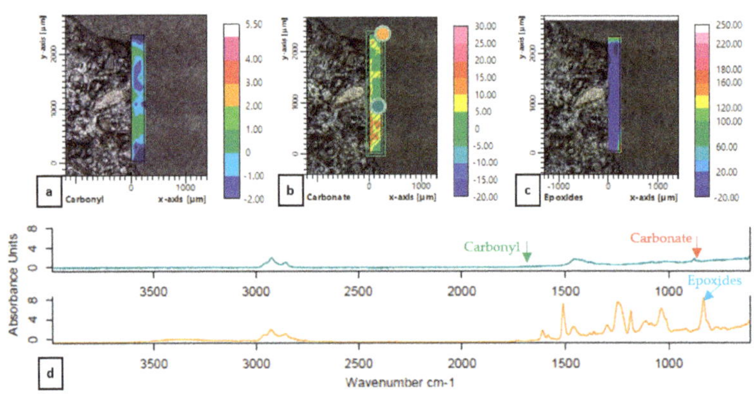

Figure 9. Chemical imaging of the GRAB sample (Area 2): carbonyl group (**a**); carbonate group (**b**); epoxides (**c**); spectral visualization of the selected points marked by colored dots in Figure 9b (**d**).

The third area is illustrated in Figure 10. A profile area of the bituminous coating of sample GRAB is presented here with the aim to track the mobilization in full depth towards the stone. The spatial resolution of Area 3 is 2800 × 610 μm^2. Similar levels of carbonyls can be observed between Areas 1 and 3 (Figures 8a and 10a). On the other hand, the levels in this area (Figure 10a) differ from the levels in Area 2 (Figure 9a). A possible hypothesis is that a lower mixing temperature and less mixing time can lead to ineffective mixing between the neat bitumen and the RA mortar. Findings in literature support this hypothesis [36]. Another observation that strengthens this hypothesis is the presence of lower levels of carbonyls and the absence of high carbonate groups in Area 2 compared to area 3 (Figure 9a, Figure 10a, Figure 9b, and Figure 10b, correspondingly) which is an indication of a low degree of blending. The chemical imaging of alkyl levels of Area 3 is demonstrated in Figure 10c, which shows that Area 3 is a bituminous zone.

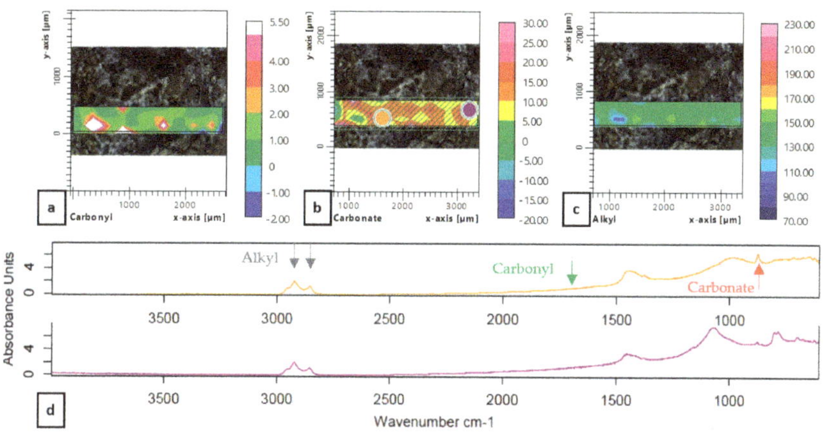

Figure 10. Chemical imaging of the GRAB sample (Area 3): carbonyl group (**a**); carbonate group (**b**); alkyl group (**c**); spectral visualization of the selected points marked by colored dots in Figure 10b (**d**).

The findings of the last two samples, MRAB and GRAR, are presented in Figures 11 and 12, respectively. The spatial resolution of the studied areas are 1000 × 340 μm^2 and 1580 × 1360 μm^2, accordingly. Concerning the carbonyl level of Area 4, a stable concentration in an order of magnitude of 2 (Figure 11a) can be observed. Since the fabrication conditions of that sample are different compared to GRAB sample, i.e., mixed at a higher temperature for a longer period of time, one can assume a higher degree of blending, as well as a more homogenized dispersion of the carbonyls within the analyzed area. Figure 12 provides information regarding the chemical imaging of the GRAR sample where a rejuvenation agent was sprayed on the RA sample. The carbonyl levels of Area 5 range from 0 to 3 (Figure 12a), which are lower compared to the concentration of carbonyl of the GRA reference sample (Figure 8a). The presence of rejuvenator can be verified by the chemical imaging in Figure 12c, where the carboxylic acids show values between 6 and 16, and also the absence of the same band in non-rejuvenated samples such as the MRAB sample (Figure 11c), where the absorption is below 0.

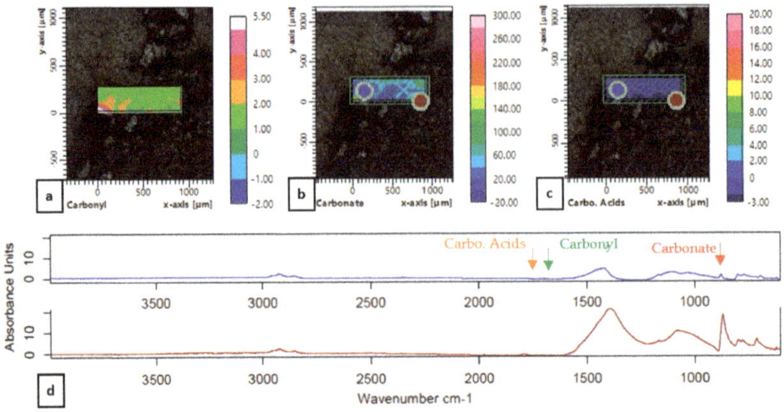

Figure 11. Chemical imaging of the MRAB sample (Area 4): carbonyl group (**a**); carbonate group (**b**); carboxylic acids (**c**); spectral visualization of the selected points marked by colored dots in Figure 11b,c (**d**).

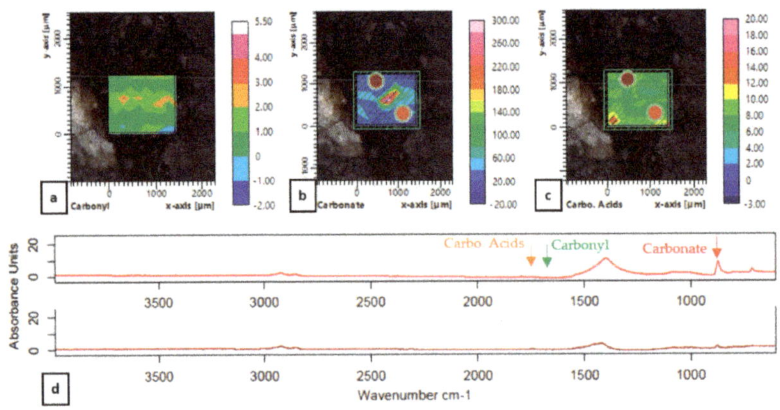

Figure 12. Chemical imaging of the GRAR sample (Area 5): carbonyl group (**a**); carbonate group (**b**); carboxylic acids (**c**); spectral visualization of the selected points marked by colored dots in Figure 12b,c (**d**).

3.2. Carbonyl Distribution as Indicator of Mobilization

The carbonyl level can be used as an indicator for ageing of the bituminous zones and therefore also as an indication of blending. The quantified carbonyl is not always a band that belongs in a bituminous zone. For that reason, by focusing on the carbonyl levels deduced only from bituminous areas, it is possible to better understand the extent of the ageing or the rejuvenation effect on those areas. Based on the integration of typical bands, a bituminous coating can be defined as the area where the following conditions are simultaneously met:

- Epoxides lower than 0; values higher than 0 indicate a resin area.
- Alkyl group higher than 0; indication of bituminous area.
- Carbonyl higher than 0.

Based on those requirements, the distribution of the carbonyl for all the areas is plotted in Figure 13a and the percentiles are presented in Table 4. Because of the large difference between the maximum level of carbonyl (27.94) and the level at the 97.5 percentile (7.16), values above the 97.5%

limit can be considered as extreme measurements. Based on that observation, the carbonyl level, which can be considered as valid within a bituminous zone, for this research is between 0 and 7.16.

Table 4. Carbonyl levels at different percentiles of the distribution.

Percentage	Percentiles	Carbonyl Level
100.0%	maximum	27.94
97.5%	97.5 percentile	7.16
75.0%	third quartile	2.00
50.0%	second quartile (median)	1.34
25.0%	first quartile	0.67
2.5%	2.5 percentile	0.09
0.0%	minimum	0.01

In Figure 13b–f, the carbonyl distributions of the individual studied zones are shown. Considering the GRA sample as the reference sample (Figure 13b), a shift can be observed towards lower carbonyl levels for the other 4 areas, which is an indication of blending. When comparing Areas 2 and 3 of the same GRAB sample (see Figure 13c,d), the probability of a carbonyl level between 0 and 1 is higher for Area 2 (66% vs. 50%), but on the other hand, some dispersed higher carbonyl levels (between 3 and 8) are present in Area 3 but not in Area 2. Those results are in line with the chemical imaging and indicate the presence of an outer zone with low carbonyl, which can possibly be interpreted as an area of a lower degree of blending consisting mostly of neat bitumen. On the other hand, Area 3 shows similar probability, for carbonyl levels higher than 1, compared to Area 1 of the GRA sample but higher probability for the lowest carbonyl levels between 0 and 1 (Figure 13b,d, respectively). Here, we can assume that an area of partial blending is present, as well as an inactive RA bituminous zone. Therefore, the assumption of full blending is not valid for these samples.

The MRAB sample exhibits a high probability of carbonyl levels between 1 and 2. This is an indication of a higher degree of blending, since lower and higher carbonyl levels are limited. The GRAR sample shows similar probabilities as Area 3 of the GRAB sample.

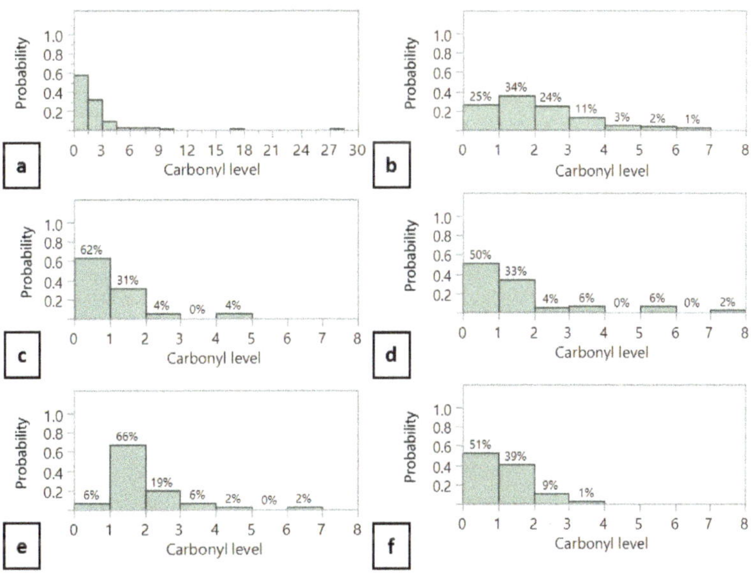

Figure 13. Probability of carbonyl levels in the bituminous zone: among all samples (**a**); GRA (**b**); GRAB Area 2 (**c**); GRAB Area 3 (**d**); MRAB (**e**); GRAR (**f**).

3.3. *Bituminous Mortar Coating of the RA Granulates*

Previous studies have discussed the mobilization of pure bitumen or mastic (bitumen and filler). Figure 14 shows part of the coating of sample MRAB. This particular sample was obtained from an actual mixture, and, as demonstrated before, for this sample, a higher degree of blending has been achieved. Therefore, it can be considered as the most appropriate representation of the result of an actual mixing procedure and thus a realistic bituminous coating of RA granulates. As can be seen in Figure 14, the surrounding coating also consists of visible fine particles larger than 63 µm (filler threshold) and smaller than 500 µm. This bituminous zone should be considered as the bituminous mortar area. Moreover, this statement is in line with the definition given for asphalt mortar in [16]. For that reason, the mobilization of mortar should be considered as a more realistic phenomenon when adding RA in bituminous mixtures rather than the mobilization of bitumen or mastic.

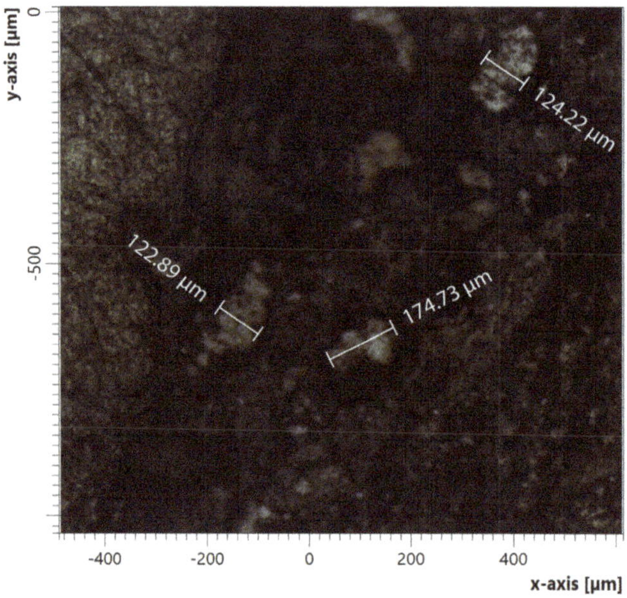

Figure 14. Bituminous mortar as coating of a RA granulate.

4. Conclusions

In this study, the application of FTIR microscopy in ATR mode on RA granulates was evaluated. A specific methodology for the preparation and the analysis of coated granulates was proposed. The samples engineered for this research were suitable for microscopic analysis, and by utilizing characteristic infrared bands, it was possible to trace zones of different composition and intensity.

The conclusions of this research are as follows:

- It is possible to trace carbonyls as an ageing indicator on actual RA granulates.
- The blending between RA mortar and neat bitumen can be characterized as only partial.
- Shorter mixing times and lower temperatures (2 min and 150 °C) can lead to large inactive RA mortar areas and a lower degree of blending.
- Higher temperature and longer mixing (5 min and 180 °C) may provide a higher degree of blending.
- Rejuvenation of bituminous coating can be traced based on the proposed method.

- The ageing index (carbonyls) for the rejuvenated area is lower than the sample areas covered with neat bitumen.

It should be noted that the described method requires an adequate amount of time for the collection of the spectra. Since this study is at a demonstrative stage, the number of samples was rather limited. Based on these results, the application of this technique is promising for future research to quantitatively evaluate the impact of different fabrication methods (mixing time and temperature) in the degree of blending, as well as its application on different and actual asphalt mixture samples.

Author Contributions: Conceptualization, A.M. and W.V.d.b.; Data Curation, A.M. and G.T.; Investigation, A.M., G.T. and G.J.; Methodology, A.M. and G.T.; Resources, W.V.d.b.; Supervision, W.V.d.b. and J.B.; Visualization, A.M. and G.J.; Writing—Original Draft, A.M.; Writing—Review and Editing, W.V.d.b., C.V. and S.T.

Funding: This research received no external funding.

Acknowledgments: The authors would like to acknowledge Patricia Dath and Pieter Ballieu from Holcim, Belgium, for their technical support and for preparation of the samples used in this study.

Conflicts of Interest: The authors declare no conflict of interest.

References

1. *NBN EN 13108-8:2016 Bituminous Mixtures—Material Specifications—Part 8: Reclaimed Asphalt*; Bureau voor Normalisatie: Brussels, Belgium, 2016.
2. Anthonissen, J. Bituminous Pavements in Flanders: Quantifying the Effect of RAP on the Environmental Impact. Ph.D. Thesis, University of Antwerp, Antwerpen, Belgium, May 2017.
3. Margaritis, A.; Van den bergh, W. Evaluation of Flemish bituminous mixtures' performance containing RAP: Statistical analysis and modelling. In Proceedings of the 13th ISAP Conference, Fortaleza, Ceará, Brazil, 19–21 June 2018.
4. Van den bergh, W.; Kara, P.; Anthonissen, J.; Margaritis, A.; Jacobs, G.; Couscheir, K. Recommendations and strategies for using reclaimed asphalt pavement in the Flemish Region based on a first life cycle assessment research. *IOP Conf. Ser. Mater. Sci. Eng.* **2017**, *236*, 012088. [CrossRef]
5. McDaniel, R.S.; Anderson, R.M. *Recommended Use of Reclaimed Asphalt Pavement in the Superpave Mix Design Method: Technician's Manual*; National Research Council (US), Transportation Research Board: Washington, DC, USA, 2001.
6. Mogawer, W.; Bennert, T.; Daniel, J.S.; Bonaquist, R.; Austerman, A.; Booshehrian, A. Performance characteristics of plant produced high RAP mixtures. *Road Mater. Pavement Des.* **2012**, *13*, 183–208. [CrossRef]
7. Cavalli, M.C.; Partl, M.N.; Poulikakos, L.D. Measuring the binder film residues on black rock in mixtures with high amounts of reclaimed asphalt. *J. Clean. Prod.* **2017**, *149*, 665–672. [CrossRef]
8. Margaritis, A.; Jacobs, G.; Hasheminejad, N.; Blom, J.; Van den bergh, W. Influence of mixing procedures on the performance of asphalt mixtures with high reclaimed asphalt content. In Proceedings of the 7th International Conference on Bituminous Mixtures and Pavements, Thessaloniki, Greece, 12–14 June 2019.
9. Kriz, P.; Grant, D.L.; Veloza, B.A.; Gale, M.J.; Blahey, A.G.; Brownie, J.H.; Shirts, R.D.; Maccarrone, S. Blending and diffusion of reclaimed asphalt pavement and virgin asphalt binders. *Road Mater. Pavement Des.* **2014**, *15*, 78–112. [CrossRef]
10. Rad, F.Y.; Sefidmazgi, N.R.; Bahia, H. Application of diffusion mechanism: Degree of blending between fresh and recycled asphalt pavement binder in dynamic shear rheometer. *Transp. Res. Rec.* **2014**, *2444*, 71–77. [CrossRef]
11. Bennert, T.; Dongré, R. Backcalculation method to determine effective asphalt binder properties of recycled asphalt pavement mixtures. *Transp. Res. Rec.* **2010**, *2179*, 75–84. [CrossRef]
12. Bowers, B.F.; Huang, B.; Shu, X.; Miller, B.C. Investigation of reclaimed asphalt pavement blending efficiency through GPC and FTIR. *Constr. Build. Mater.* **2014**, *50*, 517–523. [CrossRef]
13. Xu, J.; Hao, P.; Zhang, D.; Yuan, G. Investigation of reclaimed asphalt pavement blending efficiency based on micro-mechanical properties of layered asphalt binders. *Constr. Build. Mater.* **2018**, *163*, 390–401. [CrossRef]
14. Huang, B.; Li, G.; Vukosavljevic, D.; Shu, X.; Egan, B.K. Laboratory investigation of mixing hot-mix asphalt with reclaimed asphalt pavement. *Transp. Res. Rec.* **2005**, *1929*, 37–45. [CrossRef]

15. AbuHassan, Y.; Alin, M.; Iqbal, T.; Nazzal, M.; Abbas, A.R. Effect of extraction solvents on rheological properties of recovered asphalt binders. *J. Transp. Eng. Part B Pavements* **2019**, *145*, 04018064. [CrossRef]
16. Van den bergh, W. The Effect of Ageing on the Fatigue and Healing Properties of Bituminous Mortars. Ph.D. Thesis, Delft University of Technology, Delft, The Netherlands, December 2011.
17. Marsac, P.; Piérard, N.; Porot, L.; Van den bergh, W.; Grenfell, J.; Mouillet, V.; Pouget, S.; Besamusca, J.; Farcas, F.; Gabet, T.; et al. Potential and limits of FTIR methods for reclaimed asphalt characterisation. *Mater. Struct.* **2014**, *47*, 1273–1286. [CrossRef]
18. Hofko, B.; Alavi, M.Z.; Grothe, H.; Jones, D.; Harvey, J. Repeatability and sensitivity of FTIR ATR spectral analysis methods for bituminous binders. *Mater. Struct.* **2017**, *50*, 187. [CrossRef]
19. Karlsson, R.; Isacsson, U. Application of FTIR-ATR to characterization of bitumen rejuvenator diffusion. *J. Mater. Civ. Eng.* **2003**, *15*, 157–165. [CrossRef]
20. Mikhailenko, P.; Bertron, A.; Ringot, E. Methods for analyzing the chemical mechanisms of bitumen aging and rejuvenation with FTIR spectrometry. In Proceedings of the 8th RILEM International Symposium on Testing and Characterization of Sustainable and Innovative Bituminous Materials, Ancona, Italy, 7–9 October 2015; pp. 203–214.
21. Lamontagne, J.; Durrieu, F.; Planche, J.P.; Mouillet, V.; Kister, J. Direct and continuous methodological approach to study the ageing of fossil organic material by infrared microspectrometry imaging: Application to polymer modified bitumen. *Anal. Chim. Acta* **2001**, *444*, 241–250. [CrossRef]
22. Nivitha, M.R.; Prasad, E.; Krishnan, J.M. Ageing in modified bitumen using FTIR spectroscopy. *Int. J. Pavement Eng.* **2016**, *17*, 565–577. [CrossRef]
23. Lopes, M.; Mouillet, V.; Bernucci, L.; Gabet, T. The potential of attenuated total reflection imaging in the mid-infrared for the study of recycled asphalt mixtures. *Constr. Build. Mater.* **2016**, *124*, 1120–1131. [CrossRef]
24. Vassaux, S.; Gaudefroy, V.; Boulangé, L.; Soro, L.J.; Pévère, A.; Michelet, A.; Barragan-Montero, V.; Mouillet, V. Study of remobilization phenomena at reclaimed asphalt binder/virgin binder interphases for recycled asphalt mixtures using novel microscopic methodologies. *Constr. Build. Mater.* **2018**, *165*, 846–858. [CrossRef]
25. Rinaldini, E.; Schuetz, P.; Partl, M.N.; Tebaldi, G.; Poulikakos, L.D. Investigating the blending of reclaimed asphalt with virgin materials using rheology, electron microscopy and computer tomography. *Compos. Part B Eng.* **2014**, *67*, 579–587. [CrossRef]
26. Jiang, Y.; Gu, X.; Zhou, Z.; Ni, F.; Dong, Q. Laboratory observation and evaluation of asphalt blends of reclaimed asphalt pavement binder with virgin binder using SEM/EDS. *Transp. Res. Rec.* **2018**, *2672*, 69–78. [CrossRef]
27. Nahar, S.N.; Mohajeri, M.; Schmets, A.J.M.; Scarpas, A.; van de Ven, M.F.C.; Schitter, G. First observation of blending-zone morphology at interface of reclaimed asphalt binder and virgin bitumen. *Transp. Res. Rec.* **2013**, *2370*, 1–9. [CrossRef]
28. Mohajeri, M.; Molenaar, A.A.A.; Van de Ven, M.F.C. Experimental study into the fundamental understanding of blending between reclaimed asphalt binder and virgin bitumen using nanoindentation and nano-computed tomography. *Road Mater. Pavement Des.* **2014**, *15*, 372–384. [CrossRef]
29. Lu, X.; Talon, Y.; Redelius, P. Aging of bituminous binders–laboratory tests and field data. In Proceedings of the 4th Eurasphalt Eurobitume Congress, Copenhagen, Denmark, 21–23 May 2008; pp. 1–12.
30. Erkens, S.; Porot, L.; Gläser, R.; Glover, C.J. Aging of bitumen and asphalt concrete: Comparing state of the practice and ongoing developments in the United States and Europe. In Proceedings of the Transportation Research Board 95th Annual Meeting, Washington, DC, USA, 10–14 January 2016.
31. *NBN EN 12697-11:2012 Bituminous Mixtures-Test Methods-Part 11: Determination of the Affinity between Aggregate and Bitumen*; Bureau voor Normalisatie: Brussels, Belgium, 2012.
32. *NBN EN 12697-35:2016 Bituminous Mixtures-Test Methods-Part 35: Laboratory Mixing*; Bureau voor Normalisatie: Brussels, Belgium, 2016.
33. Cong, P.; Hao, H.; Zhang, Y.; Luo, W.; Yao, D. Investigation of diffusion of rejuvenator in aged asphalt. *Int. J. Pavement Res. Technol.* **2016**, *9*, 280–288. [CrossRef]
34. Baghaee Moghaddam, T.; Baaj, H. The use of rejuvenating agents in production of recycled hot mix asphalt: A systematic review. *Constr. Build. Mater.* **2016**, *114*, 805–816. [CrossRef]

35. Bruker Optik GmbH. *OPUS/IR Reference Manual*, version 7; Bruker Optik GmbH: Billerica, MA, USA, 2014; pp. 347–350.
36. Bressi, S.; Cavalli, M.C.; Partl, M.N.; Tebaldi, G.; Dumont, A.G.; Poulikakos, L.D. Particle clustering phenomena in hot asphalt mixtures with high content of reclaimed asphalt pavements. *Constr. Build. Mater.* **2015**, *100*, 207–217. [CrossRef]

coatings

MDPI

Article

Design of Highly Active Electrodes for Hydrogen Evolution Reaction Based on Mo-Rich Alloys Electrodeposited from Ammonium Acetate Bath

Edita Vernickaitė [1], Oksana Bersirova [2], Henrikas Cesiulis [1] and Natalia Tsyntsaru [1,3,*]

[1] Department of Physical Chemistry, Vilnius University, Naugarduko str. 24, LT-03225 Vilnius, Lithuania; edita.vernickaite@chf.vu.lt (E.V.); henrikas.cesiulis@chf.vu.lt (H.C.)

[2] V.I. Vernadsky Institute of General and Inorganic Chemistry, 32-34 Acad. Palladina ave., 03680 Kiev, Ukraine; bersirova@nas.gov.ua

[3] Institute of Applied Physics, Academiei str. 5, MD-2028 Chisinau, Moldova

* Correspondence: ashra_nt@yahoo.com; Tel.: +370-672-28632

Received: 15 January 2019; Accepted: 29 January 2019; Published: 30 January 2019

Abstract: The given research was driven by prospects to design Mo-rich coatings with iron group metals electrodeposited from a highly saturated ammonium acetate bath. The obtained coatings could be employed as prominent electrodes for the hydrogen evolution reaction (HER). It was found that the Mo content in Ni–Mo alloys can be tuned from 30 to 78 at.% by decreasing the molar ratio [Ni(II)]:[Mo(VI)] in the electrolyte from 1.0 to 0.25 and increasing the cathodic current density from 30 to 100 mA/cm^2. However, dense cracks and pits are formed due to hydrogen evolution at high current densities and that diminishes the catalytic activity of the coating for HER. Accordingly, smoother and crack-free Ni–54 at.% Mo, Co–52 at.% Mo and Fe–54 at.% Mo alloys have been prepared at 30 mA/cm^2. Their catalytic behavior for HER has been investigated in a 30 wt.% NaOH solution at temperatures ranging from 25 to 65 °C. A significant improvement of electrocatalytic activity with increasing bath temperature was noticed. The results showed that the sequence of electrocatalytic activity in alkaline media decreases in the following order: Co–52 at.% Mo > Ni–54 at.% Mo > Fe–54 at.% Mo. These peculiarities might be linked with different catalytic behavior of formed intermetallics (and active sites) in electrodeposited alloys. The designed electrodeposited Mo-rich alloys have a higher catalytic activity than Mo and Pt cast metals.

Keywords: Ni–Mo; Co–Mo; Fe–Mo alloys; electrodeposition; hydrogen evolution reaction; electrocatalysis

---

1. Introduction

Hydrogen is a clean fuel and an energy carrier that can be used for energy conversion and storage and is considered as a possible substitute for fossil fuels [1]. Electrocatalytic water splitting offers an ideal approach for highly pure hydrogen production. However, despite the multitude of on-going research, the development of an optimized, cost-effective and sustainable catalyst, which possesses a high catalytic activity for hydrogen evolution reaction (HER) is still rather appealing. Commonly, the ability of a given metal to catalyze the HER is estimated based on the exchange current density (ECD), i.e., the current density in the absence of net electrolysis at zero overpotential (at formal equilibrium potential for hydrogen evolution reaction in the particular solution). It is known, that the higher the ECD, the lower the overvoltage that must be applied to create a significant current flow. Hence, elaborated electrocatalysts should manifest exchange current densities equivalent or analogous to the ECD of polycrystalline platinum (~1 × 10^{-3} A/cm^2 in alkaline electrolytes) [2].

A considerable part of research on the design of effective cathode materials for HER has been focused on Mo alloys with iron group metals (Ni, Co, Fe) due to their superior catalytic performance in alkaline media [3–5], stability at elevated temperatures [6] and reasonable corrosion and oxidation resistance [7–10]. These characteristics, combined with good electrical conductivity, easy usage and reasonable price, are attractive parameters for selecting the cathode material for water electrolysis. It was revealed that the activity of the Ni–x at.% Mo ($12 \leq x \leq 29$) alloy for HER is much higher than that of separate metallic nickel and molybdenum electrodes [11–15]. This phenomenon is attributed to the synergistic effect of Mo dispersed in the Ni matrix, which increases the real surface area of the electrode [12,16]. In addition, the enhancement of the catalytic activity for the HER of the Ni–15 at.% Mo alloy was ascribed to the modification of electron density in d-orbitals upon alloying nickel with molybdenum [17]. In other words, this model implies that some of the electrons of the iron group metal (Ni, Fe, Co) with more filled d-bands are shared with Mo having less-filled d-orbitals. This leads to maximal bond strength and stability of the intermetallic alloy phases [18,19].

Commonly, Mo alloys with iron group metals (Ni, Co, Fe) can be synthesized by applying mechanical alloying [3,20–22], powder metallurgy [11,23,24], spraying [25] and laser cladding techniques [26]. However, fabrication processes in aqueous media are often considered as simpler, cheaper and more environmentally friendly fabrication methods than those requiring sophisticated apparatus, volatile and corrosive chemicals and extra energy that must be incurred to keep the system in a liquid state. Thus, molybdenum can be successfully co-electrodeposited in the presence of iron group metal ions (Ni(II), Co(II), Fe(II)) and appropriate complexing agents from an aqueous electrolyte. It is assumed that the molybdate ions are reduced to molybdenum oxide or hydroxide, which in the presence of iron group metal (Ni, Fe or Co) species allows the formation of the corresponding binary alloy deposits. The effective Mo alloys electrodeposition with iron group metals were carried out from citrate [3,5,7,27], citrate–ammonia [10,12], citrate-gluconate [28], ammonia [29] and pyrophosphate [30,31] aqueous electrolytes.

Moreover, it was claimed that Ni–Mo electrodes show higher electrocatalytic activity than other Ni-based binary alloys such as Ni–Co, Ni–Fe, Ni–Zn and Ni–Cr [5,29]. For a given reason, the fabrication of Ni–Mo alloys possessing the highest activity for effective hydrogen production was the target for the vast research in the last decades. There are numerous reports certifying that the catalytic activity for hydrogen evolution is qualitatively proportional to the Mo content in Mo-based alloys [11,13,17]. Therefore, researchers' efforts were directed to optimize the plating bath vs. deposition conditions in order to obtain Mo-rich alloys as effective catalysts for the HER. It was shown that electrodeposition from ammonia based aqueous solutions produces Mo alloys with up to ~41 at.% of Mo [32–36]. Coatings containing more than 40 at.% Mo have been electrodeposited from ammonium-citrate solution in the presence of imidazolium-based ionic liquids as an additive [37]. Ammonia is frequently added to improve the cathode current efficiency, however, there is also some data about its effect on decreasing the Mo content in bimetallic Mo system with iron group metal (Ni, Co, Fe) [34]. Thus, despite the reduced current efficiency, typically ammonia-free electrolytes are used for the preparation of Mo-rich coatings, e.g., the Fe–Mo electrodes containing up to 59 at.% of Mo have been prepared from a pyrophosphate bath [38,39]. Binary Fe–Mo alloys with 49 at.% of Mo have been electrochemically formed from an aqueous trisodium nitrilotriacetate bath [40]. A considerable increase in Mo content, i.e., up to 70 at.%, can be caused by the addition of Mo powder to the electrolyte [41]. In addition, it has been noted that Mo content in alloys composition can be increased by carrying out the electrodeposition under the pulse current mode [42]. The highest Mo content, 74 at.%, achieved so far by induced electrodeposition in aqueous citrate electrolyte was reported for a Ni–Mo alloy [43].

Therefore, based on the mentioned above, the given research was focused on the electrodeposition of Ni-, Co- and Fe- Mo-rich alloys from a highly saturated ammonium acetate bath. The electrolyte's composition given in Reference [44] was adapted for the electrodeposition of binary Mo-containing alloys. In order to determine the influence of the nature of the iron group metal on the catalytic activity of target coatings (Ni-, Co- and Fe- Mo-rich alloys) the electrochemical conditions were tuned in such way to ensure deposition of alloys with similar content of Mo. The catalytic activity of the electrochemically fabricated Mo-based alloy electrodes for the HER was explored in a 30 wt.% NaOH solution. In order to compare the electrochemical activity with other typical electrode materials, experiments were also performed using bare platinum electrode (same geometrical area) under the same conditions.

2. Materials and Methods

2.1. *Mo-Rich Alloys Electrodeposition*

Mo-rich alloys, namely Ni–Mo, Co–Mo and Fe–Mo, were prepared from highly saturated ammonium acetate electrolytes (Table 1) based on bath composition proposed in Reference [44] for Mo films deposition. All solutions were prepared from chemicals of analytical grade (A.R.) dissolved in demineralized water. The electrodeposition of coatings was carried out at 30 °C in order to lower the viscosity of the concentrated solutions and to avoid salt precipitation. Cu rod (surface area of 1 cm^2), platinum sheet (3 × 7 cm^2) and a saturated Ag/AgCl electrode were used as a working, counter and reference electrodes, respectively. Prior to the electrodeposition, Cu rods were washed and cleaned in an ultrasonic bath for 6–7 min and etched in an HNO_3:CH_3COOH:H_3PO_4 (1:1:1) solution at 60 °C. The thickness of the prepared Mo alloy coatings with iron group metals was calculated from gravimetric and elemental analysis data. Further, the electrocatalytic activity for the HER in 30 wt.% NaOH of fabricated cathodes has been investigated.

Table 1. Composition of electrolytes for Ni–Mo (Baths No. 1–3), Co–Mo (Bath No. 4) and Fe–Mo (Bath No. 5) coatings electrodeposition.

Bath	CH_3CO_2K	$CH_3CO_2NH_4$	$(NH_4)_2MoO_4$	$NiSO_4{\cdot}7H_2O$	$CoSO_4{\cdot}7H_2O$	$FeSO_4{\cdot}7H_2O$	pH
1				0.001 M			
2				0.002 M	–	–	8.2
3	10.2 M	10.4 M	0.004 M	0.004 M			
4				–	0.002 M		8.3
5					–	0.002 M	

2.2. *Morphological and Structural Study*

The surface morphology and chemical composition of the prepared Mo-based deposits were examined with the scanning electron microscope (SEM, Hitachi TM3000, Tokyo, Japan) equipped with an INCA energy dispersive X-ray spectroscopy detector (EDS, Oxford Instruments, Buckinghamshire, UK) at an accelerating voltage of 20 kV, respectively. Based on the chemical composition of the obtained alloys, the current efficiency (CE) was calculated according to the Faradays' law:

$$\text{CE (\%)} = \frac{Fm}{It}\left[\frac{x_i n_i}{M_i} + \frac{x_{\text{Mo}} \cdot n_{\text{Mo}}}{M_{\text{Mo}}}\right] \times 100\% \quad (1)$$

where F is Faradays constant (96485 C); m is the weight of the electrodeposit (g); I is an applied current (A); t is the time of electrodeposition (s); x_i, n_i, M_i is the content (wt.%), electrons transferred per ion, and molecular weight (g/mol) of Ni, Co or Fe, respectively; x_{Mo}, n_{Mo}, M_{Mo} is the content, wt.%, electrons transferred per particular ion, respectively; molecular weight of Mo.

The structure of the electrodeposited alloys was investigated by X-ray diffraction (XRD) methods (Rigaku MiniFlex II, Tokyo, Japan). XRD patterns were produced with Cu Kα radiation (1.5406 Å) in 2θ scanning mode from 20 to 100° with a step of 0.01°.

2.3. Electrochemical Measurements

The voltammetric measurements for the evaluation of the kinetic parameters of Mo-rich alloys for the HER were performed in a 30 wt.% NaOH solution at several temperatures (25–65 °C, with the increment of 10 °C) in a thermostatic cell. A platinum wire was used as an auxiliary electrode and a saturated Ag/AgCl electrode was used as the reference electrode. All potentials are given with respect to the Ag/AgCl reference electrode. Potentiodynamic polarization hydrogen evolution curves were recorded at the sweep rate of 2 mV/s. The cathode potential was scanned from its open circuit potential (OCP) up to −1 V. Voltammetric curves were recorded using a potentiostat/galvanostat AUTOLAB equipped with GPES software (version 4.9). Extrapolation of the polarization curves obtained at different temperatures, in the coordinates lg$i$ − η to value η = 0 give the possibility to determine the ECD (i_0). The overvoltage, η, was calculated from the following equation:

$$\eta = E - E_r \quad (2)$$

$$E_r = -\left(\frac{2.3RT}{F}\right)\text{pH} \quad (3)$$

where η is an overpotential of the HER (V), E is an experimental potential value at which the reaction takes place (V); E_r is the reversible potential value calculated from the Nernst equation (V), R is the universal gas constant (8.314472 J/K mol); T is the temperature (K); F is the Faraday constant (96,485 J/mol).

For the calculation of overpotentials at temperatures other than 25 °C, the tabulated data [45] of the temperature dependence of the potential of the saturated Ag/AgCl electrode vs. the hydrogen electrode, were used.

3. Results and Discussion

3.1. Design of Mo-Rich Alloys Coatings

The high percentage of molybdenum in Mo-based alloy electrodeposits leads commonly to the growth of the ECD value but also has a positive influence on their corrosion resistance and microhardness [10]. On the other hand, the electrodeposition of coatings having a very high molybdenum content (>38 at.% of Mo) is more sensitive to side reactions, namely the evolution of hydrogen, which can lead to the appearance of a dense net of cracks, bumps and small pits that diminish practical application of such coatings for the HER. Thus, the first step of the given research was dedicated to the selection of the optimum electrochemical conditions (bath chemistry, applied current density) in order to obtain high-quality Mo-rich alloys with a reasonable deposition rate. The first investigated system was Ni–Mo (Table 1, Baths 1–3) since a high amount of publications have reported [15,46,47] that the Ni–Mo alloy is the most promising non-noble catalysts for the HER among other refractory metal-based electrodes.

Previously, it was shown that if the ratio [Ni(II)]:[Mo(VI)] is approaching 10, the amount of Mo in the Ni–Mo deposit decreases dramatically from 65 to 20 at.% [48]. Therefore, in order to obtain Mo-rich alloys, the ratio was kept at 0.25, 0.5, 1.0. Another parameter, which influences the refractory metal content in the alloys is the applied current density. Based on a preliminary study, two cathodic current densities, namely 30 and 100 mA/cm^2, have been chosen for electrodeposition of Mo-rich alloys.

Taking these parameters into account, the dependence of Mo content on the [Ni(II)]:[Mo(VI)] ratio and the cathodic current density was evaluated (Table 2). Namely, as it was anticipated, the amount of Mo in the alloys decreases from ~85 to 36 at.% as the Ni(II) increased in the bath. Here it should be mentioned, that only the content of the metallic phase was taken into account for the evaluation regardless of the ambiguous values of oxygen and other light elements detected by the EDS analysis. The highest content of molybdenum in the Ni–Mo deposits, around 85 at.%, was achieved at a [Ni(II)]/[Mo(VI)] ratio equal to 0.25 in the plating bath. This Mo amount is significantly higher in comparison with previous works reported for Ni–Mo alloys obtained from aqueous electrolytes and is close to that found for Ni–Mo alloys prepared by metallurgical [11] or mechanical alloying techniques [49]. In order to reveal the interdependencies between bath chemistry and applied current densities, partial current densities (PSD) for Ni_{PCD}, Mo_{PCD} reduction and hydrogen evolution were also evaluated based on Faraday's law (Table 2).

Table 2. Dependence of composition, morphology and partial current densities of electrodeposited Ni–Mo alloys on the [Ni(II)]:[Mo(VI)] ratio and applied cathodic current density. Molybdenum at.% content is specified on the SEM images.

[Ni(II)]: [Mo(VI)] Ratio	Applied Cathodic *j*, 30 mA/cm^2		Applied Cathodic *j*, 100 mA/cm^2	
	SEM	Partial Cathodic *j*, mA/cm^2	SEM	Partial Cathodic *j*, mA/cm^2
1	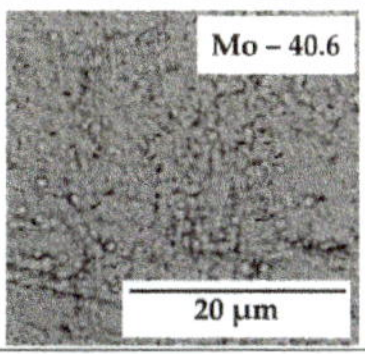	j_{Ni} = 0.6 j_{Mo} = 1.3 j_{H_2} = 28.1	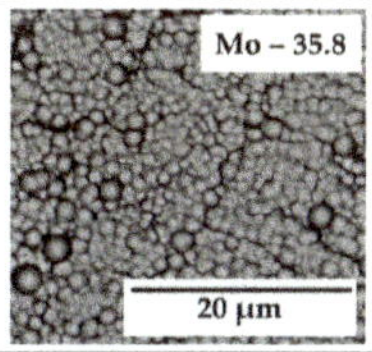	j_{Ni} = 1.3 j_{Mo} = 2.2 j_{H_2} = 96.5
0.5	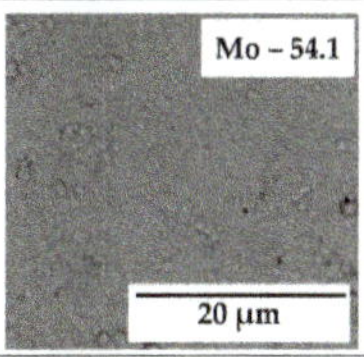	j_{Ni} = 0.3 j_{Mo} = 1.2 j_{H_2} = 28.5	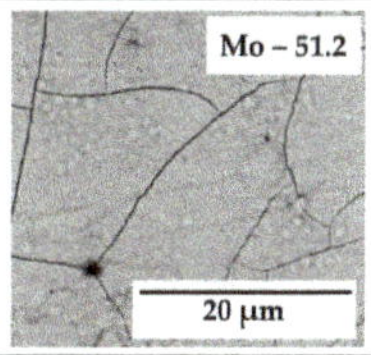	j_{Ni} = 0.6 j_{Mo} = 2.1 j_{H_2} = 97.3
0.25	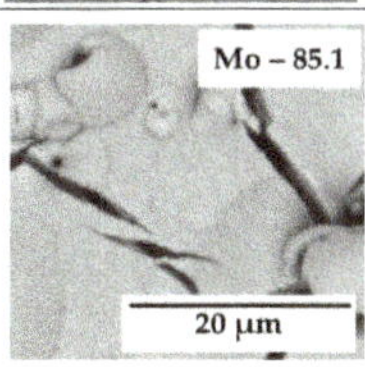	j_{Ni} = 0.02 j_{Mo} = 0.27 j_{H_2} = 29.71	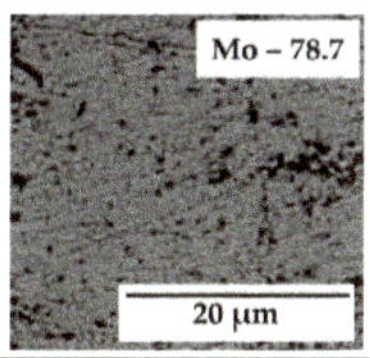	j_{Ni} = 0.2 j_{Mo} = 2.3 j_{H_2} = 97.5

Namely, the increase of the [Ni(II)]:[Mo(VI)] ratio increases the Ni_{PCD} and consequently Ni content in the deposit the higher applied current density (overpotential) accelerates the reduction of Ni(II) rather than Mo(VI) compounds. The side reaction is accelerated by a higher Mo content in the alloy that leads to the propagation of large micro-cracks, especially for Ni–Mo alloys deposited at a [Ni(II)]/[Mo(VI)] ratio $\leq$ 0.5 and having more than 50 at.% of Mo. Our results are in a good agreement with Reference [50], where it was shown that the cracks in the Ni–Mo alloys deposited from a citrate solution have been tracked at a Mo content higher than ~30 at.% but from an ammonium-citrate electrolyte [3] they appear even at lower Mo content (~21 at.%).

Hence, in order to obtain Mo-rich alloys without visible defects, the applied current density of 30 mA/cm^2 and a [Ni(II)]:[Mo(VI)] ratio of 0.5 should be viewed as the optimum conditions. This ratio allows for a four times increase in the Mo_{PCD} in comparison with the ratio of 0.25. It suggests that the electroactive complex should contain both molybdenum and nickel species. At the higher ratio (higher Ni(II) concentration), the Mo_{PCD} does not change significantly but the Mo content in the alloy decreases (from 54 to 40 at.%). Notably, at a current density of 100 mA/cm^2, the Mo_{PCD} is practically the same for all investigated [Ni(II)]:[Mo(VI)] ratios, suggesting that electroactive Mo-containing species under such conditions reaches saturation and has no significant effect on alloy electrodeposition. Furthermore, at a higher applied current density, an additional roughening due to the pronounced nodular structure of the coatings is obtained, that can be interconnected with abundant hydrogen evolution, which leads to cracks and holes on the surface (Table 2). On the one hand, in Reference [51], it was shown that the cracked surfaces are characterized by higher HER activity and ascribed to an increased surface area of the active centers due to the microcracks but on the other hand, a certain amount of the hydrogen enters the open pores (cracks) of the deposit and it starts peeling off around the pores. Hence, it is obvious that the application of such coatings in the industrial processes is not recommendable [30]. Moreover, due to high PCDs for hydrogen reduction, the current efficiency in all investigated cases is rather low (<10%). This is a common characteristic for Ni–Mo co-deposition [52] that is associated with the formation of a mixed Mo oxides layer in the presence of an excess of Mo(VI) ions in the bath, which hinders the further reduction.

Accordingly, based on the experimental results obtained for Ni–Mo alloys, the following optimal conditions were adapted for electrodeposition of Co–Mo and Fe–Mo alloys: cathodic current density 30 mA/cm^2 and [Me(II)/Mo(VI)] = 0.5. This allowed for the electrodeposition of Mo-rich coatings (Co–52 at.% Mo and Fe–54 at.% Mo) coupled with suitable morphology (crack-free coatings with a less rough globular surface) and to evaluate the influence of iron group metal on the catalytic activity for HER in alkaline media. The obtained morphology of Mo-rich alloys was quite similar regardless of the iron group metal (Figure 1).

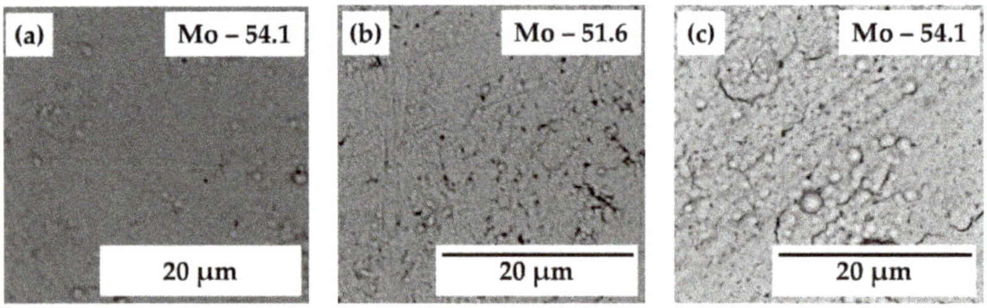

Figure 1. SEM images of electrodeposited at 30 mA/cm^2 and [Me(II)]/[MoO_4^{2-}] = 0.5 coatings: Ni–Mo (**a**), Co–Mo (**b**) and Fe–Mo (**c**). The time of electrolysis was 1 h and the thickness of all deposits was ~10 μm.

The structure and crystallite size was evaluated by XRD analysis. A characteristic broad peak at 2θ = 43°–44° was obtained for Ni–54 at.% Mo and is depicted in Figure 2a. According to the thermodynamic equilibrium data, the solubility limit of Mo in the fcc Ni structure at room temperature is ~17 at.%. When the Mo content exceeds this limit, an amorphous microstructure can be noticed and the formation of intermetallic Ni_4Mo, Ni_3Mo, NiMo compounds becomes possible for Ni–Mo alloys having >25 at.% of Mo [31,53]; a line with a solid solution of Ni in Mo. It is also known that the broadening of the XRD peak is related to the refinement of crystallite size that typically occurs with an increasing Mo content [31]. According to the literature, the mean crystallite size of Ni–Mo coatings can decrease from 50 to 2 nm by increasing the Mo content from 1 to 38 at.%, respectively [35,37,54]. This corresponds to a crystallite size of the investigated Ni–Mo coatings as small as ~2 nm. Notably, Ni–Mo deposits consisting of such small crystallites can have a lower overpotential for hydrogen evolution due to the larger concentration of crystal lattice defects and dislocations, which are considered as active centers for HER [12,19].

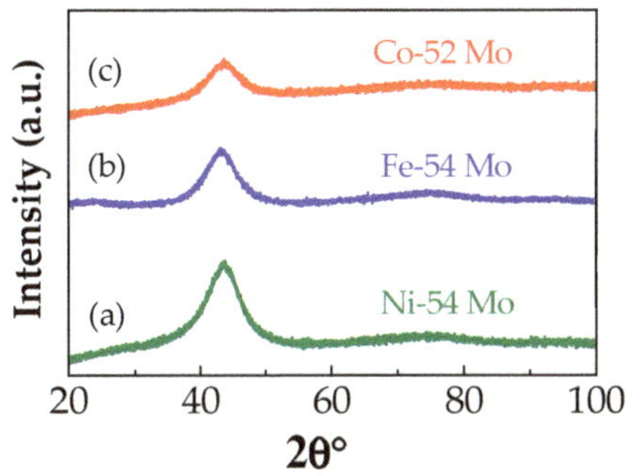

Figure 2. XRD pattern for: (**a**) Ni–54 Mo, (**b**) Fe–54 Mo and (**c**) Co–52 Mo electrodeposits (composition is given in at.%).

The XRD patterns for Fe–54 at.% Mo and Co–52 at.% Mo alloys showed the analogous crystalline structure to the Ni–54 at.% Mo coating (Figure 2b,c). Since the presence of one broad peak in the X-ray diffraction patterns makes it difficult to interpret the results, it can be only proposed that a mixture of Mo solid solution in the iron group metal and corresponding intermetallic compounds were formed. For the Fe–Mo (also Fe–W) system, the Mössbauer spectroscopy supports this presumption and suggests that deposits having more than 17 at.% of refractory metal consists of a mixture of molybdenum solid solution in α-Fe and intermetallic phases, e.g., Fe_3Mo, Fe_2Mo [55,56]. In the case of the Co–Mo alloy, some of the studies reported that a Mo solid solution in cobalt and intermetallic Co_3Mo is formed [36,57].

The different intermetallic phases should have an impact on their activity for the HER. It was emphasized that the maximum electrocatalytic activity could be achieved for intermetallic phases of highest symmetry and minimal entropy, such as Laves phases or A_3B types (Co_3Mo, Ni_3Mo, Fe_3Mo) and the Brewer theory for intermetallic bonding predicts as the most stable systems [58]. Thus, it was reported that films consisting of a Co_3Mo phase (for Co–x at.% Mo, $18 \leq x \leq 28$) have the best electrocatalytic properties among other Co–Mo alloys having lower molybdenum contents [59]. Furthermore, in Reference [60], it was concluded that Co_3Mo intermetallic compounds are more stable in a hot alkaline solution than other Co–Mo phases. Similarly, in the case of the Fe–Mo alloy system, the lowest overvoltage for hydrogen evolution at a current density of 200 mA/cm^2 has been observed for the Fe–47 at.% Mo sample with a predominant Fe_3Mo intermetallic compound phase [38].

3.2. Catalytic Behavior

The electrocatalytic activity for hydrogen evolution in 30 wt.% NaOH of Ni–Mo, Co–Mo and Fe–Mo samples containing ~52–54 at.% of Mo in their composition was evaluated using a linear scan voltammetry method that allows for the determination of the apparent exchange current densities. Polarization curves and semi-logarithmic coordinates of all chosen systems obtained at 25 °C are presented in Figure 3. In order to compare the catalytic behavior of Mo-rich Ni-, Co-, Fe-Mo coatings, the cast Mo and Pt electrodes were used. Table 3 summarizes the calculated apparent exchange current densities (i_0), the overpotentials at a selected current density of 200 mA/cm^2 ($\eta_{0.2}$) and the current densities obtained at an overpotential of 0.3 V ($i_{0.3}$); calculated Tafel slopes (b_c).

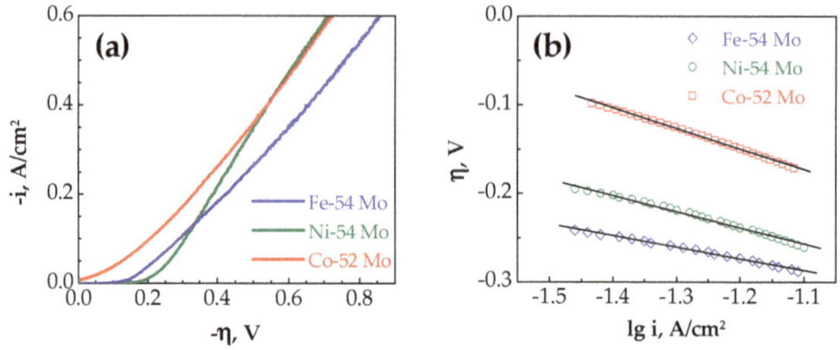

Figure 3. Cathodic polarization curves of (**a**) Mo-rich electrodeposits in 30 wt.% KOH at 25°C and (**b**) plots in semi-logarithmic coordinates. The scan rate was 2 mV/s (composition is given in at.%).

Table 3. Calculated apparent exchange current densities (i_0), Tafel slopes (b_c), overpotentials (η_i) at i = 0.3 A/cm^2 and current densities (i_η) at η = −0.3 V for hydrogen evolution different electrodes (composition is given in at.%).

Parameter	Electrode				
	Ni–54 Mo	Co–52 Mo	Fe–54 Mo	Mo	Pt
i_o (mA/cm^2)	0.62	1.90	0.23	2.90×10^{-5}	2.63
b_c (mV/dec)	128	132	152	231	122
η_i (V)	0.46	0.43	0.54	0.67	0.48
i_η (A/cm^2)	9.1×10^{-2}	1.8×10^{-1}	9.3×10^{-2}	5.4×10^{-4}	8.2×10^{-2}

Binary Mo alloy cathodes possess 10^5 times higher apparent exchange current densities than cast Mo, thus are more active for the HER. However, all alloys demonstrated lower apparent exchange current densities for the HER at 25 °C in comparison with a bare Pt electrode. However, the overpotentials required to obtain current densities of 300 mA/cm^2 and current densities at −0.3 V for active bimetallic Mo alloys ware similar to those determined for Pt, making them competitive electrodes for hydrogen production. Notably, as it was mentioned above, the nature of the iron group metal affects the catalytic activity for the HER and the cathodic current density for the Co–52 at.% Mo coating is higher in comparison to that of Ni–54 at.% and Fe–54 at.% Mo, thus indicating the best catalytic performance among the synthesized electrodes, which is consistent with the lowest $\eta_{0.2}$ and $i_{0.3}$ values.

It is well known that the lower Tafel slope implies a lower electrochemical electrode polarization during the HER process, particularly at a high current density. The values of the Tafel slope for Ni–54 at.% Mo and Co–52 at.% Mo deposits under high polarization conditions are 128 mV/s and 132 mV/dec, respectively. Meanwhile, the Tafel slope for Fe–54 at.% Mo under these conditions shifted to more positive values, i.e., increased up to 152 mV/dec and it can be related to the higher iron affinity to the air and the presence of a thin oxide film on the surface that is characterized by a lower conductivity that impedes the electron transfer rate [15].

An improvement of catalytic activity for the HER with the operation temperature, as is desired for practical industrial alkaline electrolysis has been reported [36,42]. Accordingly, in the present study, the electrodeposited alloy electrodes were tested at temperatures ranging from 25 to 65 °C by applying 10 °C increments. A general comparison of the performance of electrodeposited Ni–54 at.% Mo, Fe–54 at.% Mo, Co–52 at.% Mo and cast separate metals in the temperature range 25–65 °C is given in Figure 4. The apparent exchange current densities were calculated from the linear region at low overpotential values and are presented in Table 4. As it was expected, the electrocatalytic activity of Ni–54 at.% Mo, Fe–54 at.% Mo and Co–52 at.% Mo is significantly higher than cast Mo in the whole tested temperature range. Moreover, the results suggest that at an elevated temperature (>35 °C) Mo-rich alloys have a more prominent HER outperformance than Pt investigated in our laboratory. Although the HER activity, namely i_0, was found to increase with temperature, the Tafel slopes for the Mo-based alloys remained almost constant and varied in the range of 120–150 mV/dec. This phenomenon has been discussed in terms of the entropic contribution towards free energy of activation [61]. Among all investigated systems, the Co–52 at.% Mo electrode demonstrates the best performance towards the HER, particularly at temperatures higher than 45 °C. These results correspond well with findings published in Reference [62] where it was confirmed that Co–Mo co-deposits are characterized by a higher catalytic activity and stability in alkaline water electrolysis than Ni–Mo, Co–W and Ni–W alloy electrodes. Moreover, in Reference [63], it was shown that catalytic activity depends on the metal-hydrogen bond strength and absorption sites in the alloy available to hydrogen and thus the electrochemically charged H content decreases in the series of Co–Mo > Co–W > Ni–Mo. Furthermore, comparing Co–Mo, Co–W and Ni–Mo electrodeposits, the thermal desorption of hydrogen occurs at the lowest temperature on Co–Mo showing a faster recombination step of H atoms possible on this alloy, thus improving its electrocatalytic performance [63].

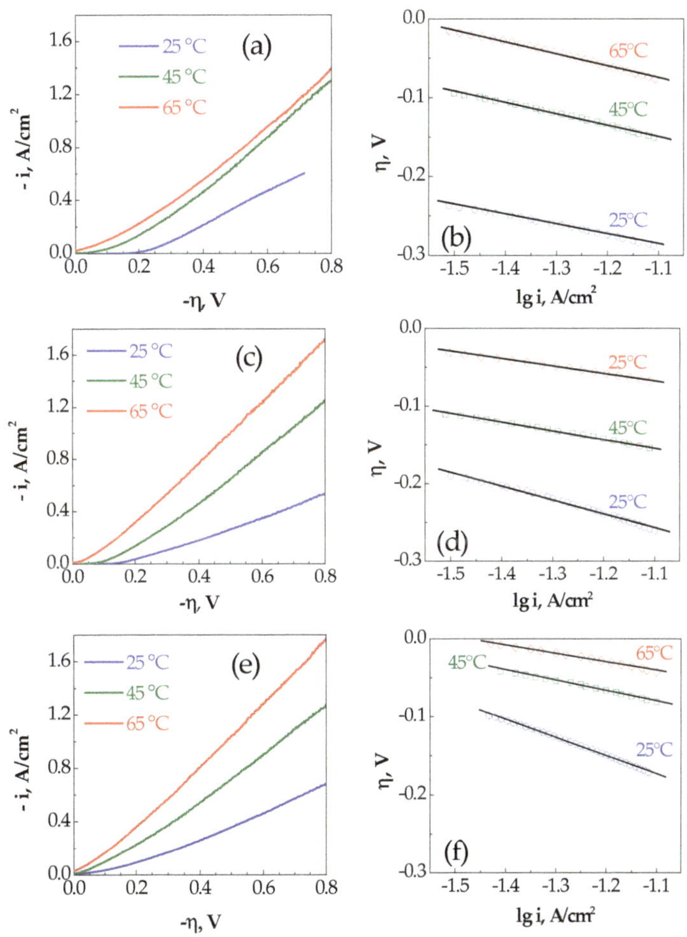

Figure 4. Effect of temperature on cathodic polarization curves at Ni–54 Mo (**a**), Fe–54 Mo (**c**), Co–52 Mo (**e**) electrodeposits in 30 wt.% KOH at different temperatures and plots in semi-logarithmic coordinates for Ni–54 Mo (**b**), Fe–54 Mo (**d**), Co–52 Mo (**f**) electrodes. The scan rate was 2 mV/s (composition is given in at.%).

Table 4. Experimental values of the apparent exchange current (i_0, in mA/cm^2) and Tafel slopes (b_c, in mV/dec) for Ni–54 Mo, Co–52 Mo and Fe–54 Mo alloy electrodes at different temperatures (composition is given in at.%).

Sample	Measurement temperature (°C)									
	25		35		45		55		65	
	i_0	b_c	i_0	b_c	i_0	b_c	i_0	b_c	i_0	b_c
Ni-54 Mo	0.62	128	3.21	128	7.32	119	1.47	116	25.4	123
Fe-54 Mo	0.23	152	0.99	148	4.33	142	5.83	139	14.6	145
Co-52 Mo	1.90	132	9.53	130	17.1	128	32.0	121	46.2	119
Mo	2.9×10^{-2}	231	8.3×10^{-2}	220	1.9×10^{-1}	220	2.1×10^{-1}	215	2.3×10^{-1}	221
Pt	2.63	122	3.68	129	6.51	120	2.63	125	11.5	125

In many previous publications the Ni–Mo alloy coatings are characterized by a lower overpotential value, as compared to the Co–Mo [14]. Though, there is also some information suggesting that Co–Mo coatings have a better catalytic activity for the HER compared to Ni–Mo deposits [62,64]. At the first glance, these controversial results could be attributed to the different alloy preparation techniques that are capable of yielding an uneven composition, morphology and structure of the prepared samples and thus, directly influence the catalytic properties of the samples. However, the clear tendency between the nature of the iron group metal effect during alloying with Mo and catalytic behavior also cannot be easily defined even for the Ni–Mo, Co–Mo and Fe–Mo cathodes fabricated using the same electrodeposition technique (Table 5). As it can be seen, the exchange current densities, even for alloys having a similar chemical composition, vary depending on the selected alkaline media and temperature.

In general, the present study shows the catalytic activity for the Ni–Mo alloy is comparable to previously reported samples under similar experimental conditions (Figure 5). Moreover, in the case of the Co–Mo and Fe–Mo system, the apparent exchange current density calculated in our work is significantly higher than it was expected from other authors observations (the corresponding columns are not given in the figure due to a significantly lower value). This could be attributed to the more concentrated alkaline media and higher temperature used in the present study.

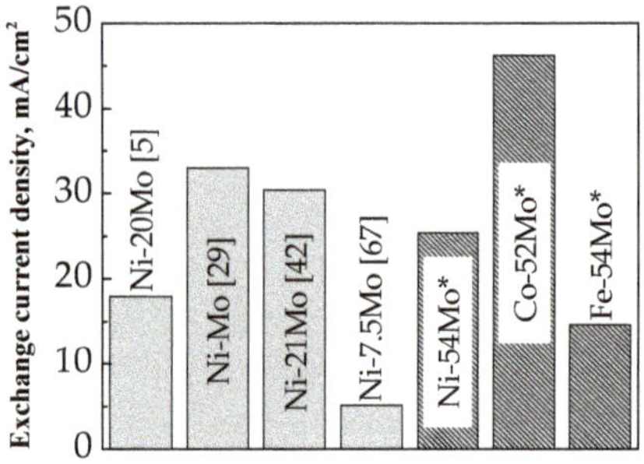

Figure 5. The comparison of apparent exchange current densities (i_0) towards the HER on Mo alloys with iron group metals determined in this study (*) with the published data. The experiments were performed in NaOH at 60–65 °C. The composition of alloys is given in at.%.

Notably, the Fe–Mo coating demonstrates lower exchange current densities for the HER in an alkaline environment among other electrodes investigated in this study. This may be related to the higher iron affinity to the air by forming an oxide, hydroxide, or mixed film that physically separates the metal surface from the electrolyte.

Table 5. Comparison of the HER apparent exchange current density values, i_0, extracted from the published data for electrodeposited Ni–Mo, Co–Mo and Fe–Mo alloy electrodes.

Sample	Mo content (at.%)	Media	i_0 (mA/cm^2)	Reference
Ni–Mo	26	8.25 M NaOH, 85 °C	44.4	[4]
	20	6 M KOH; 80 °C	18.62	[5]
	33.8	7 M KOH, 25 °C	2.8	[19]
	–	7 M KOH, 80 °C	55.24	[29]
	25	2 M NaOH, 30 °C	3.1×10^{-2}	[32]
	20.8	11 M NaOH; 80 °C	42.4	[42]
	29.8	1 M NaOH, 30 °C	11.1	[65]
	27.5	1 M KOH	3.18×10^{-3}	[66]
	7.5	6 M KOH; 70 °C	7.3	[67]
NiMo-modified Ni foam	2.5	0.1 M NaOH, 25 °C	4.1×10^{-2}	[28]
Ni + Mo composite	44	5 M KOH; 25 °C	1.0	[68]
Ni–Mo–rGO	30.8	1 M KOH; 25 °C	4.31×10^{-3}	[69]
Ni–Mo	10.4	7 M KOH, 25 °C	2.6×10^{-2}	[62]
Co–Mo	21.4		2.3×10^{-2}	
Co–Mo	40.9	1 M NaOH	1.5	[33]
	32	0.5 M NaOH; 60 °C	6.9×10^{-3}	[36]
	25	1 M NaOH, 25 °C	0.13	[59]
	19	1 M KOH, 25 °C	0.36	[70]
	33	1 M NaOH, 30 °C	5.0×10^{-2}	[71]
Fe–Mo	59.3	1 M NaOH, 25 °C	2.4×10^{-3}	[39]

In order to obtain a more complete picture of electrocatalytic behavior for the HER, the corresponding E_a values for all tested systems have been calculated considering the linear dependence $\lg(i_0) = f(1/T)$ using Arrhenius equation:

$$E_a = -2.303R\frac{\partial(\lg i_0)}{\partial(1/T)} \quad (4)$$

where E_a is the activation energy (J/mol), i_0—apparent exchange current density, T—temperature (K).

Figure 6 shows the Arrhenius plots for the as-deposited Ni–54 at.% Mo, Co–52 at.% Mo and Fe–54 at.% Mo electrodes. From the slope of these plots E_a values of 36.6, 32.5 and 27.9 kJ/mol for Fe–Mo, Ni–Mo and Co–Mo electrodes were determined, respectively. It is known that the lower the E_a value is, the lower the energy requirements for hydrogen production. Thus, it is obvious that the charge transfer rate is favored by Co–52 at.% Mo alloys electrodeposition, since this electrode showed slightly lower activation energy than that of the Ni–54 at.% Mo and Fe–54 at.% Mo coatings.

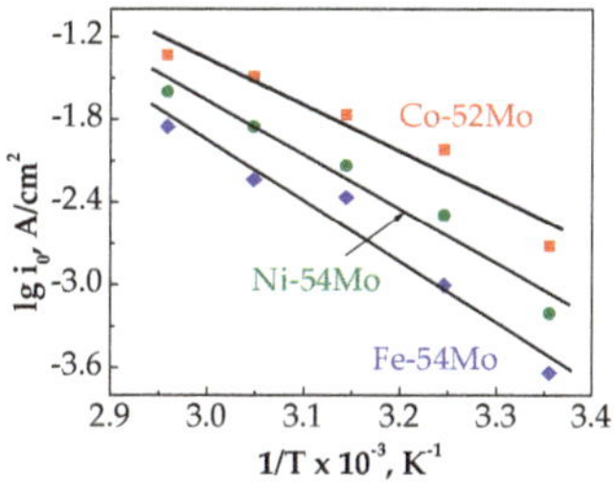

Figure 6. Arrhenius plots for Ni–54 Mo, Co–52 Mo and Fe–54 Mo electrocatalysts (composition is given in at.%).

4. Conclusions

- The electrodeposition of Mo-rich (36–82 at.%) alloys with iron group metals (Ni, Co, Fe) from highly saturated ammonium acetate aqueous electrolytes is reported. The composition was affected by the [Ni(II)]/[Mo(VI)] ratio in the plating bath and cathodic current density.
- The electro-catalytic activity towards cathodic hydrogen evolution in 30 wt.% NaOH solution in the temperature range of 25–65 °C on the electrodeposited Ni–54 at.% Mo, Co–52 at.% Mo, Fe–54 at.% Mo and Co–52 at.% Mo alloy coatings characterized by amorphous-like structure has been investigated.
- Bimetallic Mo-based alloys are considered as more active for the HER in comparison with the cast Mo and Pt since they demonstrate higher apparent exchange current densities in the tested temperature range. The apparent exchange current density of hydrogen for Co–52 at.% Mo deposits were considerably higher than those for Ni–54 at.% Mo and Fe–54 at.% Mo alloy coatings and this can be attributed to the formation of stable intermetallic Co_3Mo phase which ensures optimal Co and Mo distribution over the surface and produces larger active sites for the HER.
- The calculated activation energy values suggest that the Mo alloy coating with iron group metals shows promising electrocatalytic activity for the HER and among all investigated samples, the Co–52 at.% Mo electrode is characterized by a lower activation energy (27.9 kJ/mol) than the Ni–54 at.% Mo (32.5 kJ/mol) and Fe–54 at.% Mo (36.6 kJ/mol) coatings.

Author Contributions: Investigation, E.V. and O.B.; Methodology, E.V. and O.B.; Supervision, H.C. and N.T.; Visualization, E.V.; Writing–Original Draft Preparation, E.V.; Writing–Review & Editing, E.V., H.C. and N.T.

Funding: This research has received funding from Horizon 2020 research and innovation program under MSCA-RISE-2017 (No. 778357) and from Research Lithuanian Council project (No 09.3.3-LMT-K-712-08-0003).

Conflicts of Interest: The authors declare no conflict of interest.

References

1. Studer, S.; Stucki, S.; Speight, J.D. Hydrogen as a Fuel. In *Hydrogen as a Future Energy Carrier*; Zttel, A., Borgschulte, A., Schlapbach, L., Eds.; Wiley-VCH Verlag GmbH & Co. KGaA: Weinheim, Germany, 2008; pp. 23–69.
2. Sheng, W.; Gasteiger, H.A.; Shao-Horn, Y. Hydrogen oxidation and evolution reaction kinetics on platinum: Acid vs alkaline electrolytes. *J. Electrochem. Soc.* **2010**, *157*, B1529–B1536. [CrossRef]
3. Halim, J.; Abdel-Karim, R.; El-Raghy, S.; Nabil, M.; Waheed, A. Electrodeposition and characterization of nanocrystalline Ni–Mo catalysts for hydrogen production. *J. Nanomater.* **2012**, *2012*, 845673. [CrossRef]
4. Aaboubi, O. Hydrogen evolution activity of Ni–Mo coating electrodeposited under magnetic field control. *Int. J. Hydrogen Energy* **2011**, *36*, 4702–4709. [CrossRef]
5. Raj, I.A.; Venkatesan, V.K. Characterization of nickel-molybdenum and nickel-molybdenum-iron alloy coatings as cathodes for alkaline water electrolysers. *Int. J. Hydrogen Energy* **1988**, *13*, 215–223.
6. Kapoor, G.; Huang, Y.; Sarma, V.S.; Langdon, T.G.; Gubicza, J. Influence of Mo alloying on the thermal stability and hardness of ultrafine-grained Ni processed by high-pressure torsion. *J. Mater. Res. Technol.* **2017**, *6*, 361–368. [CrossRef]
7. Feng, C.; Qian, W.; Liu, J.; Han, S.; Fu, N.; Ye, F.; Lin, H.; Jiang, J. Effect of ultrasonication on Ni–Mo coatings produced by DC electroformation. *RSC Adv.* **2016**, *6*, 30652–30660. [CrossRef]
8. Laszczyńska, A.; Tylus, W.; Winiarski, J.; Szczygieł, I. Evolution of corrosion resistance and passive film properties of Ni–Mo alloy coatings during exposure to 0.5 M NaCl solution. *Surf. Coat. Technol.* **2017**, *317*, 26–37. [CrossRef]
9. Chassaing, E.; Portail, N.; Levy, A.-F.; Wang, G. Characterisation of electrodeposited nanocrystalline Ni–Mo alloys. *J. Appl. Electrochem.* **2004**, *34*, 1085–1091. [CrossRef]
10. Huang, P.-C.; Hou, K.-H.; Wang, G.-L.; Chen, M.-L.; Wang, J.-R. Corrosion resistance of the Ni–Mo alloy coatings related to coating's electroplating parameters. *Int. J. Electrochem. Sci.* **2015**, *10*, 4972–4984.

11. Jakšić, J.M.; Vojnović, M.V.; Krstajić, N.V. Kinetic analysis of hydrogen evolution at Ni–Mo alloy electrodes. *Electrochim. Acta* **2000**, *45*, 4151–4158. [CrossRef]
12. Xu, C.; Zhou, J.; Zeng, M.; Fu, X.; Liu, X.; Li, J. Electrodeposition mechanism and characterization of Ni–Mo alloy and its electrocatalytic performance for hydrogen evolution. *Int. J. Hydrogen Energy* **2016**, *41*, 13341–13349. [CrossRef]
13. Mech, K.; Zabinski, P.; Mucha, M.; Kowalik, R. Electrodeposition of catalytically active Ni–Mo alloys/elektroosadzanie aktywnych katalitycznie stopów Ni–Mo. *Arch. Metall. Mater.* **2013**, *58*, 227–229. [CrossRef]
14. Jeremiasse, A.W.; Bergsma, J.; Kleijn, J.M.; Saakes, M.; Buisman, C.J.N.; Cohen Stuart, M.; Hamelers, H.V.M. Performance of metal alloys as hydrogen evolution reaction catalysts in a microbial electrolysis cell. *Int. J. Hydrogen Energy* **2011**, *36*, 10482–10489. [CrossRef]
15. Navarro-Flores, E.; Chong, Z.; Omanovic, S. Characterization of Ni, NiMo, NiW and NiFe electroactive coatings as electrocatalysts for hydrogen evolution in an acidic medium. *J. Mol. Catal. A Chem.* **2005**, *226*, 179–197. [CrossRef]
16. Manazoğlu, M.; Hapçı, G.; Orhan, G. Electrochemical deposition and characterization of Ni–Mo alloys as cathode for alkaline water electrolysis. *J. Mater. Eng. Perform.* **2016**, *25*, 130–137. [CrossRef]
17. Martinez, S.; Metikoš-Huković, M.; Valek, L. Electrocatalytic properties of electrodeposited Ni–15Mo cathodes for the HER in acid solutions: Synergistic electronic effect. *J. Mol. Catal. A Chem.* **2006**, *245*, 114–121. [CrossRef]
18. Lu, G.; Evans, P.; Zangari, G. Electrocatalytic properties of Ni-based alloys toward hydrogen evolution reaction in acid media. *J. Electrochem. Soc.* **2003**, *150*, A551–A557. [CrossRef]
19. Huang, L.; Yang, F.; Xu, S.; Zhou, S. Studies of structure and electrocatalytic hydrogen evolution on electrodeposited nanocrystalline Ni–Mo alloy electrodes. *Trans. IMF* **2001**, *79*, 136–139. [CrossRef]
20. Kedzierzawski, P.; Oleszak, D.; Janik-Czachor, M. Hydrogen evolution on hot and cold consolidated Ni–Mo alloys produced by mechanical alloying. *Mater. Sci. Eng. A* **2001**, *300*, 105–112. [CrossRef]
21. González, G.; Sagarzazu, A.; Villalba, R.; Ochoa, J. Comparative study of NiW, NiMo and MoW prepared by mechanical alloying. *J. Alloys Compd.* **2007**, *434–435*, 525–529. [CrossRef]
22. Schulz, R.; Huot, J.Y.; Trudeau, M.L.; Dignard-Bailey, L.; Yan, Z.H.; Jin, S.; Lamarre, A.; Ghali, E.; Van Neste, A. Nanocrystalline Ni–Mo alloys and their application in electrocatalysis. *J. Mater. Res.* **1994**, *9*, 2998–3008. [CrossRef]
23. Yang, C.; Muránsky, O.; Zhu, H.; Thorogood, G.J.; Huang, H.; Zhou, X. On the origin of strengthening mechanisms in Ni–Mo alloys prepared via powder metallurgy. *Mater. Des.* **2017**, *113*, 223–231. [CrossRef]
24. Bhattacharjee, P.P.; Ray, R.K.; Upadhyaya, A. Development of cube texture in pure Ni, Ni–W and Ni–Mo alloys prepared by the powder metallurgy route. *Scripta Materialia* **2005**, *53*, 1477–1481. [CrossRef]
25. Tang, X.; Xiao, L.; Yang, C.; Lu, J.; Zhuang, L. Noble fabrication of Ni–Mo cathode for alkaline water electrolysis and alkaline polymer electrolyte water electrolysis. *Int. J. Hydrogen Energy* **2014**, *39*, 3055–3060. [CrossRef]
26. Li, X.; Liu, Z.; Wang, Y. Microstructure and corrosion properties of laser cladding MoNi based alloy coatings. *Sci. China Technol. Sci.* **2014**, *57*, 980–989. [CrossRef]
27. Aaboubi, O.; Chopart, J.-P. Magnetic field effect on molybdenum based alloys electrodeposition. *ECS Trans.* **2010**, *25*, 27–34.
28. Mikolajczyk, T.; Pierozynski, B. Influence of electrodeposited Ni–Mo alloy on hydrogen evolution reaction at nickel foam cathode. *Int. J. Electrochem. Sc.* **2018**, *13*, 621–630. [CrossRef]
29. González-Buch, C.; Herraiz-Cardona, I.; Ortega, E.M.; García-Antón, J.; Pérez-Herranz, V. Development of Ni–Mo, Ni–W and Ni–Co macroporous materials for hydrogen evolution reaction. *Chem. Eng. Trans.* **2013**, *32*, 865–870.
30. Krstajic, N.V.; Jovic, V.D.; Gajic-krstajic, L.; Jovic, B.M.; Antozzi, A.L.; Martelli, G.N. Electrodeposition of Ni–Mo alloy coatings and their characterization as cathodes for hydrogen evolution in sodium hydroxide solution. *Int. J. Hydrogen Energy* **2008**, *33*, 3676–3687. [CrossRef]
31. Donten, M.; Cesiulis, H.; Stojek, Z. Electrodeposition of amorphous/nanocrystalline and polycrystalline Ni–Mo alloys from pyrophosphate baths. *Electrochim. Acta* **2005**, *50*, 1405–1412. [CrossRef]
32. Gennero de Chialvo, M.R.; Chialvo, A.C. Hydrogen evolution reaction on smooth Ni$(1-x)$ + Mo(x) alloys $(0 \leq x \leq 0.25)$. *J. Electroanal. Chem.* **1998**, *448*, 87–93. [CrossRef]

33. Kuznetsov, V.V.; Kalinkina, A.A.; Pshenichkina, T.V.; Balabaev, V.V. Electrocatalytic properties of cobalt-molybdenum alloy deposits in the hydrogen evolution reaction. *Russ. J. Electrochem.* **2008**, *44*, 1350–1358. [CrossRef]
34. Podlaha, E.J.; Landolt, D. Induced codeposition: 1. An experimental investigation of Ni–Mo alloys. *J. Electrochem. Soc.* **1996**, *143*, 885–892. [CrossRef]
35. Bigos, A.; Beltowska-Lehman, E.; Kot, M. Studies on electrochemical deposition and physicochemical properties of nanocrystalline Ni–Mo alloys. *Surf. Coat. Technol.* **2017**, *317*, 103–109. [CrossRef]
36. Casciano, P.N.S.; Benevides, R.L.; Santana, R.A.C.; Correia, A.N.; de Lima-Neto, P. Factorial design in the electrodeposition of Co–Mo coatings and their evaluations for hydrogen evolution reaction. *J. Alloys Compd.* **2017**, *723*, 164–171. [CrossRef]
37. Allahyarzadeh, M.H.; Roozbehani, B.; Ashrafi, A.; Shadizadeh, S.R.; Kheradmand, E. Electrochemically deposition of high Mo content amorphous/nanocrystalline Ni–Mo using ionic liquids as additive. *ECS Trans.* **2012**, *41*, 11–28.
38. Elezović, N.; Grgur, N.B.; Krstajić, N.V.; Jović, V.D. Electrodeposition and characterization of Fe–Mo alloys as cathodes for hydrogen evolution in the process of chlorate production. *J. Serb. Chem. Soc.* **2005**, *70*, 879–889.
39. Elezović, N.R.; Jović, V.D.; Krstajić, N.V. Kinetics of the hydrogen evolution reaction on Fe–Mo film deposited on mild steel support in alkaline solution. *Electrochim. Acta* **2005**, *50*, 5594–5601. [CrossRef]
40. Barbano, E.P.; de Carvalho, M.F.; Carlos, I.A. Electrodeposition and characterization of binary Fe–Mo alloys from trisodium nitrilotriacetate bath. *J. Electroanal. Chem.* **2016**, *775*, 146–156. [CrossRef]
41. Niedbała, J. Production of Ni − Mo + Mo composite coatings with increased content of embeded Mo. *Arch. Mater. Sci.* **2006**, *27*, 121–127.
42. Han, Q.; Cui, S.; Pu, N.; Chen, J.; Liu, K.; Wei, X. A study on pulse plating amorphous Ni–Mo alloy coating used as HER cathode in alkaline medium. *Int. J. Hydrogen Energy* **2010**, *35*, 5194–5201. [CrossRef]
43. Sun, S.; Podlaha, E.J. Electrodeposition of Mo-Rich, MoNi alloys from an aqueous electrolyte. *J. Electrochem. Soc.* **2012**, *159*, D97–D102. [CrossRef]
44. Morley, T.J.; Penner, L.; Schaffer, P.; Ruth, T.J.; Bénard, F.; Asselin, E. The deposition of smooth metallic molybdenum from aqueous electrolytes containing molybdate ions. *Electrochem. Commun.* **2012**, *15*, 78–80. [CrossRef]
45. Kahlert, H. Reference electrodes. In *Electroanalytical Methods*, 2nd ed.; Scholz, F., Ed.; Springer: Berlin/Heidelberg, Germany, 2010; pp. 291–308.
46. Fosdick, S.E.; Berglund, S.P.; Mullins, C.B.; Crooks, R.M. Evaluating electrocatalysts for the hydrogen evolution reaction using bipolar electrode arrays: Bi- and trimetallic combinations of Co, Fe, Ni, Mo, and W. *ACS Catal.* **2014**, *4*, 1332–1339. [CrossRef]
47. Raj, I.A.; Vasu, K.I. Transition metal-based hydrogen electrodes in alkaline solution? electrocatalysis on nickel based binary alloy coatings. *J. Appl. Electrochem.* **1990**, *20*, 32–38. [CrossRef]
48. Sanches, L.S.; Domingues, S.H.; Marino, C.E.B.; Mascaro, L.H. Characterisation of electrochemically deposited Ni–Mo alloy coatings. *Electrochem. Commun.* **2004**, *6*, 543–548. [CrossRef]
49. Rodríguez-Valdez, L.; Estrada-Guel, I.; Almeraya-Calderon, F.; Neri-Flores, M.A.; Martinez-Villafane, A.; Martinez-Sanchez, R. Electrochemical performance of hydrogen evolution reaction of Ni–Mo electrodes obtained by mechanical alloying. *Int. J. Hydrogen Energy* **2004**, *29*, 1141–1145. [CrossRef]
50. Beltowska-Lehman, E. Kinetics of induced electrodeposition of alloys containing Mo from citrate solutions. *Phys. Status Solidi C* **2008**, *5*, 3514–3517. [CrossRef]
51. Benaicha, M.; Allam, M.; Dakhouche, A.; Hamla, M. Electrodeposition and characterization of W-rich NiW alloys from citrate electrolyte. *Int. J. Electrochem. Sci.* **2016**, *11*, 7605–7620. [CrossRef]
52. Bigos, A.; Bełtowska-Lehman, E.; Kania, B.; Szczerba, M. Ni–Mo alloys electrodeposited under direct current from citrate-ammonia plating bath. *Inżynieria Materiałowa.* **2013**, *34*, 135–139.
53. Costovici, S.; Manea, A.-C.; Visan, T.; Anicai, L. Investigation of Ni–Mo and Co–Mo alloys electrodeposition involving choline chloride based ionic liquids. *Electrochim. Acta* **2016**, *207*, 97–111. [CrossRef]
54. Karolus, M.; Łągiewka, E. Crystallite size and lattice strain in nanocrystalline Ni–Mo alloys studied by Rietveld refinement. *J. Alloys Compd.* **2004**, *367*, 235–238. [CrossRef]
55. Kuznetsov, V.V.; Golyanin, K.E.; Ladygina, Y.S.; Pshenichkina, T.V.; Lyakhov, B.F.; Pokholok, K.V. Electrodeposition of iron–molybdenum alloy from ammonium–citrate solutions and properties of produced materials. *Russ. J. Electrochem.* **2015**, *51*, 748–757. [CrossRef]

56. Nicolenco, A.; Tsyntsaru, N.; Fornell, J.; Pellicer, E.; Reklaitis, J.; Baltrunas, D.; Cesiulis, H.; Sort, J. Mapping of magnetic and mechanical properties of Fe-W alloys electrodeposited from Fe(III)-based glycolate-citrate bath. *Mater. Des.* **2018**, *139*, 429–438. [CrossRef]
57. Kinh, V.Q.; Chassaing, E.; Saurat, M. Electroplating of crack-free corrosion resistant Co–Mo alloy coatings. *Electrodepos. Surf. Treat.* **1975**, *3*, 205–212. [CrossRef]
58. Jakšić, M.M. Advances in electrocatalysis for hydrogen evolution in the light of the Brewer-Engel valence-bond theory. *J. Mol. Catal.* **1986**, *38*, 161–202. [CrossRef]
59. Zhou, Q.F.; Lu, L.Y.; Yu, L.N.; Xu, X.G.; Jiang, Y. Multifunctional Co–Mo films fabricated by electrochemical deposition. *Electrochim. Acta* **2013**, *106*, 258–263. [CrossRef]
60. Lee, C.R.; Kang, S.G. Electrochemical stability of Co–Mo intermetallic compound electrodes for hydrogen oxidation reaction in hot KOH solution. *J. Power Sources* **2000**, *87*, 64–68. [CrossRef]
61. Conway, B.E.; Tessier, D.F.; Wilkinson, D.P. Temperature dependence of the Tafel slope and electrochemical barrier symmetry factor. *J. Electrochem. Soc.* **1989**, *136*, 2486–2493. [CrossRef]
62. Fan, C.; Piron, D.L.; Sleb, A.; Paradis, P. Study of electrodeposited nickel-molybdenum, nickel-tungsten, cobalt-molybdenum, and cobalt-tungsten as hydrogen electrodes in alkaline water electrolysis. *J. Electrochem. Soc.* **1994**, *141*, 382–387. [CrossRef]
63. Lupu, D.; Mărginean, P.; Biriş, A.R. Hydrogen in some synergetic electrocatalysts. *J. Alloys Compd.* **1996**, *245*, 146–152. [CrossRef]
64. Domínguez-Crespo, M.A.; Plata-Torres, M.; Torres-Huerta, A.M.; Arce-Estrada, E.M.; Hallen-López, J.M. Kinetic study of hydrogen evolution reaction on Ni_{30} Mo_{70}, $Co_{30}Mo_{70}$, $Co_{30}Ni_{70}$ and $Co_{10}Ni_{20}Mo_{70}$ alloy electrodes. *Mater. Charact.* **2005**, *55*, 83–91. [CrossRef]
65. Manazoğlu, M.; Hapçı, G.; Orhan, G. Effect of electrolysis parameters of Ni–Mo alloy on the electrocatalytic activity for hydrogen evaluation and their stability in alkali medium. *J. Appl. Electrochem.* **2016**, *46*, 191–204. [CrossRef]
66. Shetty, S.; Mohamed Jaffer Sadiq, M.; Bhat, D.K.; Hegde, A.C. Electrodeposition and characterization of Ni–Mo alloy as an electrocatalyst for alkaline water electrolysis. *J. Electroanal. Chem.* **2017**, *796*, 57–65. [CrossRef]
67. Tasic, G.S.; Maslovara, S.P.; Zugic, D.L.; Maksic, A.D.; Marceta Kaninski, M.P. Characterization of the Ni–Mo catalyst formed in situ during hydrogen generation from alkaline water electrolysis. *Int. J. Hydrogen Energy* **2011**, *36*, 11588–11595. [CrossRef]
68. Panek, J.; Budniok, A. Ni + Mo composite coatings for hydrogen evolution reaction. *Surf. Interface Anal.* **2008**, *40*, 237–241. [CrossRef]
69. Shetty, S.; Sadiq, M.M.J.; Bhat, D.K.; Hegde, A.C. Electrodeposition of Ni–Mo–rGO composite electrodes for efficient hydrogen production in an alkaline medium. *New J. Chem.* **2018**, *42*, 4661–4669. [CrossRef]
70. Kublanovsky, V.S.; Yapontseva, Y.S. Electrocatalytic properties of Co–Mo alloys electrodeposited from a citrate-pyrophosphate electrolyte. *Electrocatalysis* **2014**, *5*, 372–378. [CrossRef]
71. Subramania, A.; Sathiyapriya, A.; Muralidharan, V. Electrocatalytic cobalt–molybdenum alloy deposits. *Int. J. Hydrogen Energy* **2007**, *32*, 2843–2847. [CrossRef]

coatings

MDPI

Review

State of the Art in Directed Energy Deposition: From Additive Manufacturing to Materials Design

Adrita Dass [1] and Atieh Moridi [2,*]

1 Materials Science and Engineering Department, Cornell University, Ithaca, NY 14850, USA
2 Mechanical and Aerospace Engineering Department, Cornell University, Ithaca, NY 14850, USA
* Correspondence: moridi@cornell.edu

Received: 11 June 2019; Accepted: 24 June 2019; Published: 29 June 2019

Abstract: Additive manufacturing (AM) is a new paradigm for the design and production of high-performance components for aerospace, medical, energy, and automotive applications. This review will exclusively cover directed energy deposition (DED)-AM, with a focus on the deposition of powder-feed based metal and alloy systems. This paper provides a comprehensive review on the classification of DED systems, process variables, process physics, modelling efforts, common defects, mechanical properties of DED parts, and quality control methods. To provide a practical framework to print different materials using DED, a process map using the linear heat input and powder feed rate as variables is constructed. Based on the process map, three different areas that are not optimized for DED are identified. These areas correspond to the formation of a lack of fusion, keyholing, and mixed mode porosity in the printed parts. In the final part of the paper, emerging applications of DED from repairing damaged parts to bulk combinatorial alloys design are discussed. This paper concludes with recommendations for future research in order to transform the technology from "form" to "function," which can provide significant potential benefits to different industries.

Keywords: additive manufacturing; directed energy deposition; process maps; laser engineered net shaping; dilution; solidification cooling rate; process-microstructure relationship

1. Introduction

Additive manufacturing (AM), also popularly known as 3D printing, is at the frontier of development for manufacturing diverse parts and has also been referred to as the third industrial revolution [1–3]. AM has an advantage over other conventional manufacturing techniques, making it possible to print complex shapes without the need for several conventional processing steps, such as expensive tooling, dies, or casting molds [3–5]. There are several reviews on AM covering different aspects from process dynamics to post-processing [6–10]. This paper focuses on one such type of AM process, popularly known as directed energy deposition (DED), (also more specifically as laser engineered net shaping (LENS™), blown powder additive manufacturing, laser metal deposition system, and directed laser deposition), which has attracted significant attention due to its ability to print metals and potentially any metal-alloy system, notably functionally gradient materials [11,12]. Another important application utilizing DED is the remanufacturing or repairing of a component to increase its lifespan and hence reduce environmental impact [12,13].

DED systems have a concentrated energy source and a stream of raw material, both intersecting at a common focal point, generally in the presence of an inert shield gas. The energy density generated at a particular point melts the raw materials in and around that spot, giving rise to melt pool formation. There are various types of DED systems, which include (but are not limited to): Powder-feed and wire-feed based DED (on the basis of the type of feedstock), melt based DED, and kinetic energy based DED (on the basis of the type of energy source). Melt based DED could be further subclassified as laser based DED, electron-beam based DED, plasma based DED, and electric arc based DED. The powder based DED system has been studied extensively in the literature and is the most commonly used metal DED technique. It predominantly uses a laser beam as the heat source. Wire based DED processes provide a lower resolution as compared to laser-beam powder based processes, but have a higher deposition rate and the ability to build larger structures [14,15]. They generally use an electron-beam, plasma, or electric arc as the heat source. Electron-beam based AM (EBAM), which has a high energy focused electron beam in vacuum, can fuse almost any metal. EBAM is commercialized by Sciaky, Inc. and it is mainly used for manufacturing near net shape parts [16]. Plasma based AM uses a controlled plasma source to melt the metal particles. This is a relatively new AM technology and is commercially being used by Norsk Titanium to build mainly titanium parts [17]. Electric arc based DED melts the wire feed to deposit the layers. Emerging technology, like metal big area additive manufacturing (mBAAM) [14], takes advantage of the principle of electric arc welding to print big parts. Kinetic energy based DED systems, often referred to as Cold Spray, use a converging-diverging nozzle to accelerate micron sized particles to supersonic velocities [18]. Beyond a critical impact velocity, micron sized particles adhesively bond to the substrate and build up material in the form of a coating as well free standing bulk components [19]. Figure 1 summarizes the different DED categories in the form of a flowchart.

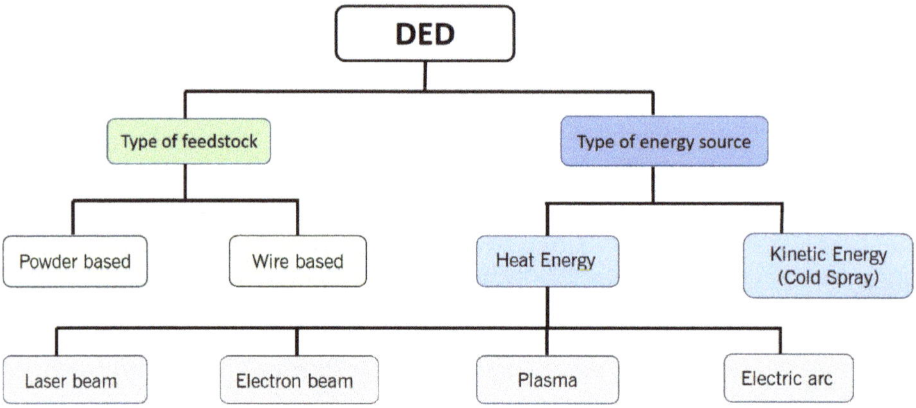

Figure 1. Classification of Directed Energy Deposition (DED) systems.

This paper will focus on the powder based DED systems in detail. It covers sections on metal and alloy systems, emphasizing the potential of DED, modelling efforts and process variables (surface tension, Marangoni effect, dimensionless numbers, energy distribution in a DED process, process-microstructure relationships, dilution), common defects (porosity, solute segregation and changes in chemical compositions, printability of alloys), mechanical properties (tensile strength, hardness, fatigue, residual stress), DED process control and monitoring, determination of optimal processing parameters by establishing process maps and the regions where high amount of defects are expected, application, and emerging technologies (DED of metal parts in biomedical applications, welding and cladding, repair, bulk combinatorial alloy design, construction materials, and hybrid AM). The paper will conclude with an overview of possible future perspectives of the field.

Metal and Alloy Systems

DED has been used to print mainly functionally graded materials, metal-matrix composites, and coatings. Each system was developed for a targeted application (e.g., enhancing biocompatibility, improving oxidation resistance, mechanical, and tribological properties, interfacial strength, etc.). Table 1 summarizes selected material systems with an emphasis on the potential applications.

Table 1. Selected studies showing DED as an emerging method to print complex metal and alloy systems.

Material System	Reported Functionality or Application or Process Optimization as Applicable *	References
Ni coated TiC on Inconel 625	Improved mechanical properties	[20]
Invar + TiC	Invar has a very low coefficient of thermal expansion which was used to make parts with low thermal stresses	[21]
Ti–48Al–2Cr–2Nb + TiC	Used to study optimum process parameters, high temperature structural applications and it exhibited twice the hardness of Ti-6Al-4V	[22]
TiC/Ti	Combines the high temperature and wear resistance of ceramics with good mechanical properties of metals	[23]
VC on SS304	Ultrahigh temperature coating on stainless steel to improve high temperature performance	[24]
Ti-6Al-4V-Al_2O_3	Conventionally processed ceramics requires post-processing (e.g., high temperature sintering), but this compositionally graded ceramic deposit on Ti-6Al-4V (having compositions, like pure Ti-6Al-4V, Ti-6Al-4V + Al_2O_3, and pure Al_2O_3) can be achieved in one step using computer aided manufacturing, thereby reducing the need for post-processing	[25]
Ti-TiO_2	TiO_2 on the surface of porous Ti increases surface wettability and biocompatibility	[26]
Compositionally graded alumina on SS303	Better interfacial properties of coating due to growth of Al_2O_3 coarse columnar microstructure in the direction of deposition	[27]
CoCrMo on porous Ti-6Al-4V	Porous implant that can eliminate stress shielding issue associated with fully dense implants and CoCrMo coting can improve the biocompatibility	[28]
Ti + SiC	Good electrochemical and tribological properties, non-toxic for biological implants	[29,30]
Calcium phosphate on Ti	Load bearing implants with high hardness and wear resistance	[31]
CNT (carbon nanotubes) + calcium phosphate + Ti-6Al-4V	Tribo-film formation improved wear resistance for Ti-6Al-4V, CNT facilitated in-situ carbide formation enhancing hardness	[32]
Ti-Zr-BN	Demonstrated superior mechanical properties with respect to pure Ti; could be used to improve surface and bulk properties.	[33]
Hydroxyapatite (HA) coating on Ti-6Al-4V	Deposition of Ti/HA interfacial layer on Ti-6Al-4V using LENS™, followed by plasma spraying of HA, to improve bond strength and increase osteoconductivity of metallic implants	[34]
Ni-18Al-11Cr-9C and Ni-14Al-8Cr-29C	Improved wear resistance	[35]
YSZ (yttria stabilized zirconia) on Ni based superalloy	Good thermal barrier protection and good thermal cycling resistance	[36]
ZTA (zirconia toughened alumina)	ZTA is tougher than pure alumina, making it useful in many industries, but it exhibits poor surface quality and therefore was post-processed using rotary ultrasonic machining, which combines both grinding and ultrasonic machining processes	[37]

Table 1. *Cont.*

Material System	Reported Functionality or Application or Process Optimization as Applicable *	References
Inconel and ceramic powders	Optimized the process parameters with the help of mode FRONTIER® software that helped minimize energy consumption and material wastes, and maximizes powder melting	[38]
Ti-Si-N coatings on Ti	Higher hardness and wear resistant coatings on Ti (with varying ratios of Ti-Si, higher Ti content lead to higher hardness)	[39]
Lead zirconate titanate (PZT)	Reasonable dielectric properties achieved with DED without post-processing, can be used potentially in the PZT embedded sensors and transducers on structural materials	[40]
Alumina (Al_2O_3)	Easier manufacturing of refractory materials by DED compared to expensive and difficult traditional processing techniques	[41–45]
Al_2O_3 + YAG (yttrium aluminum garnet)	This eutectic ceramic with water-cooled substrate showed a 10% increase in microhardness and a fracture toughness increase of 8.5%, compared to non-water cooled substrate	[46]
Tri calcium phosphate (TCP) ceramic	Good biocompatibility with cell differentiation ability for load bearing bone implants	[47]
Al_2O_3 + ZrO_2	DED with ultrasonic vibrations helped in achieving crack free parts and good mechanical properties	[48,49]
Al_2O_3-ZrO_2 (Y_2O_3) eutectic ceramic structures	Direct fabrication by DED without binders, having acceptable mechanical performance	[50]
Ni-Ti	Influence of laser power on the properties of DED fabricated equi-atomic Ni-Ti composition and the effect of laser power on corrosion resistance	[51]
SS316L	Greater resistance to corrosion after heat treatment	[52]
Nb-Ti-Si-Cr	High temperature structural strength and oxidation resistance	[53]
Fe-Cr-Ni-Mo-W-B	Optimized volume of hard precipitates using CALPHAD, achieved crack free deposition on carbon steel substrate when it was preheated to 400 °C	[54]

* This table shows potential for printing diverse combinations of materials with DED, rather than looking at all applications or functionalities in great detail.

2. DED Process Variables and Modelling Efforts

2.1. Overview of Powder-Fed DED Process Physics and Thermal History

DED is a non-equilibrium processing technique, which has very fast cooling rates, often on the order of 10^3 to 10^5 K/s [55,56] for laser and electron beam energy sources. Major process parameters for laser based DED include: Laser power, laser beam spot size, powder or wire feed rate, scanning speed, carrier gas flow rate, clad angle, feedstock properties, and layer dimensions. Therefore, a diverse set of processing parameters coupled with the complex transport phenomena, including conduction of heat into the substrate, convection due to Marangoni effects, and radiation accompanied by the shield gas, lead to a difficulty in understanding the effect of these individual process parameters on the overall DED process. Figure 2 schematically shows the complex thermal history during the multi-layer DED process, and the trend of an increasing peak temperature with an increasing layer number due to the accumulation of heat in the system [57].

Modelling efforts are beneficial to complement experimental data. The model should be close to the real DED system, taking into account the transient temperature and heat flow, complex transport phenomena, heating and cooling cycles, solidification rate, etc. These in turn give important information about the microstructure, defects, texture, and mechanical properties [57]. Table 2 summarizes some important modelling efforts of the DED process in the literature.

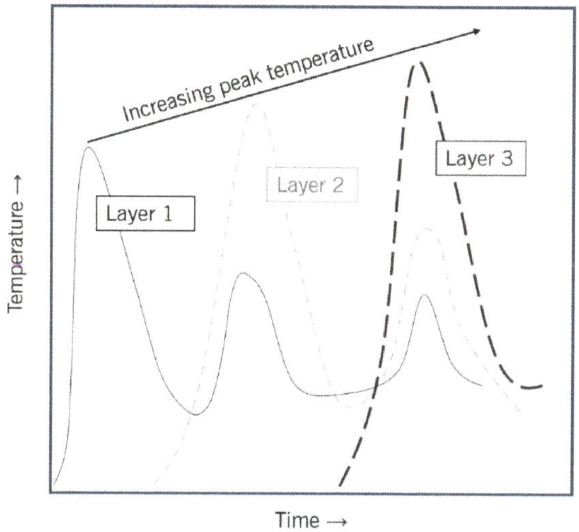

Figure 2. Schematic showing the thermal cycles for three consecutive layers during DED, and the corresponding peak temperatures for each layer.

Table 2. Selected studies on the modelling efforts of various DED processes.

Model	Explanation	Reference
Powder stream mass and temperature of the melt pool	Mass flow model taking into account powder flow into the melt pool using negative enthalpy method (subtracting the mass out of the model compared to the one considering no mass).	[58]
Powder flow trajectory simulations	The powder catchment efficiency was dependent on the carrier gas flow rate, with a higher flow rate giving more catchment; bigger powders had a smaller velocity due to inertia, leading to less catchment efficiency. Smaller powders also had less catchment efficiency.	[59]
Forced convection events	This model takes into account forced convection effects due to inert gas flow inside the chamber, and also the powder flow along with the inert gas, to reduce errors in simulations.	[60]
Powder nozzle physics	Numerical simulations to study the powder flow and determine the contribution of nozzle geometry in affecting the powder flow.	[61]
Columnar to equiaxed transition	Thermal modelling and computational thermodynamics to study microstructure heterogeneity during AM.	[62]
Phase transformations in Ti-6Al-4V	Large scale phase field modelling approach to predict the localized phase transformations in Ti-6Al-4V.	[63]
Heat transfer and cooling rates during DED	Modelling of melt pool temperatures, velocities, cyclic thermal cooling behavior, and peak temperatures.	[64–68]
Residual stress evolution	Simulations of localized residual stresses using finite element analysis, and assuming quasi-static models.	[69,70]

2.2. *Surface Tension and Marangoni Effect*

The Marangoni effect is a convective heat transfer phenomenon, which affects the melt pool flow dynamics and indirectly contributes towards porosity, which is a major concern in DED processed materials. In 1982, Heiple and Roper [71] postulated the theory that Marangoni forces generated due to the differences in surface tension and temperatures along the melt pool lead to more spattering and circulation of the liquid melt pool. The movement is mainly from regions of high surface tension to low surface tension (γ), finally leading to variable melt pool penetration. The strength of the Marangoni flow for any DED process can be determined through the dimensionless Marangoni number (M_a) [72], as provided in Equation (1):

$$M_a = \frac{d\gamma}{dT}\frac{dT}{dx}\frac{L^2}{\eta\alpha} \quad (1)$$

where γ is the surface tension, dT/dx is the temperature gradient, α is the thermal diffusivity, L is the characteristic length, and η is the viscosity of the melt pool. The surface tension gradient (the slope of the graph) qualitatively governs the melt pool movement. Figure 3 schematically shows how the variation of the surface tension with temperature affects the melt pool geometry. Figure 3a shows how the melt pool length is small due to a negative surface tension gradient and signifies bulk turbulence flow in the melt pool. Figure 3b shows how the melt pool depth increases with a positive gradient of surface tension and surface turbulence occurs in the melt pool, which could also potentially trap undesired oxides in the bulk. Figure 3c shows the transition from a positive to negative surface tension gradient at a certain temperature, T_o. This transition also indicates a melt pool flow transition from surface turbulence to bulk turbulence. The surface tension and therefore the internal melt pool flow could be controlled, to a certain extent, using surface active elements. For example, in an Fe system, changes in the concentration of the surface-active elements, like sulphur and oxygen, were shown to modify the internal melt pool flow [72,73].

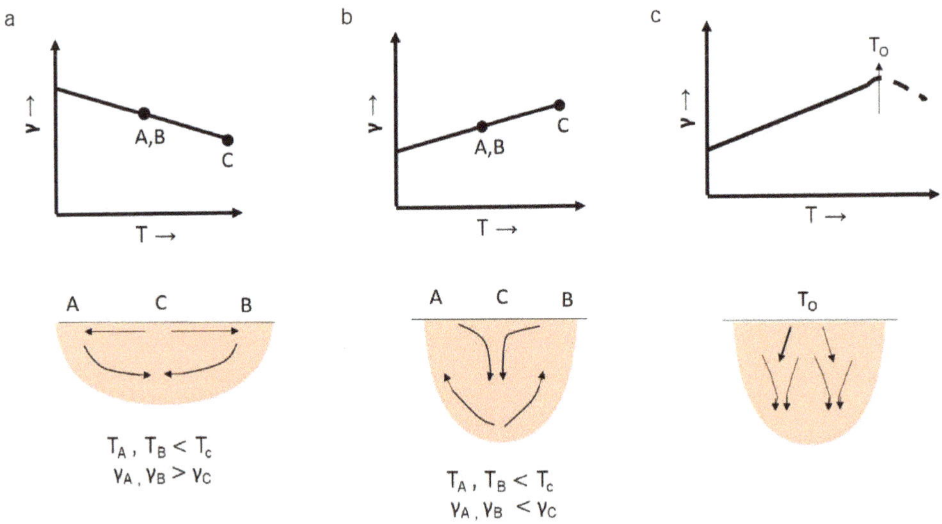

Figure 3. Schematic of Marangoni effect using the Heiple–Roper theory of weld pool geometry, depending on the surface tension and temperature of the melt pool, as applicable for DED systems. (**a**) Melt pool geometry when the surface tension gradient is negative, (**b**) melt pool geometry when the surface tension gradient is positive, and (**c**) melt pool geometry when the surface tension gradient shifts from positive to negative.

Besides the Marangoni flow force, other forces, like aerodynamic drag (outward drag forces caused by the plume formed above the melt pool), buoyancy (upward movements of the melt pool due to density changes caused by thermal gradients inside the melt pool), electromagnetic, and Lorentz forces (forces due to electric and magnetic fields generated by the source), may also be present during the DED process [72].

2.3. Dimensionless Numbers

Other than the Marangoni number, there are several other dimensionless numbers which enable capturing of the accumulative effect of various process parameters [74]. While these dimensionless numbers are not specifically developed for DED, they can capture the DED process variable relationships very well. Table 3 summarizes three such dimensionless numbers for laser based DED systems (these can be extended to other heat sources as well).

Table 3. Dimensionless numbers and their definitions.

Name	Formula *	Definition	Literature Reported
Non-dimensional Heat Input	$Q^* = \frac{P/V}{P_r/V_r}$ where P is the laser power, V is the scanning speed, P_r is the reference laser power, and V_r is the reference scanning speed (P_r and V_r provide the lowest heat input per unit length of deposited material)	Energy per unit length of material used to deposit layers	[74]
Peclet Number	$P_e = \frac{UL}{\alpha}$ where U is the characteristic velocity, α is the thermal diffusivity of the alloy, and L is the characteristic length.	The relative strength of convection to conduction in the system	[75]
Fourier Number	$F_0 = \frac{\alpha\tau}{L^2}$ where α is the thermal diffusivity, τ is the characteristic time scale, and L is the characteristic length	Gives relative value between heat dissipation and heat storage rates	[76]

* Characteristic length refers to either the thickness of the layers, the melt pool width or depth, and the laser beam spot size; characteristic velocity is considered to be the maximum velocity of the molten metal; characteristic time is defined as the length of the melt pool divided by the scanning speed.

2.4. Energy Distribution in a DED System

During DED, there should be conservation of mass, momentum, and energy, like any other physical system. Therefore, keeping track of how the initial heat source energy is distributed during the process will be beneficial in further improving the DED process to reduce energy loss and maximize energy for melting powders. Calorimetric measurements of the energy absorbed by the substrate (Q_{ABS}), energy absorbed by the powder (Q_{DEP}), energy reflected by the substrate (Q_{REF}), and energy lost by the powder (Q_{LOST}) due to evaporation and lack of fusion for DED of Ti-6Al-4V and Inconel were performed in the literature [77]. The important parameter is the bulk absorption coefficient (β), which gives the ratio of the energy for a particular component of energy (Q_{ABS}, Q_{DEP}, Q_{REF}, or Q_{LOST}) with respect to the total energy supplied to the system (Q_{IN}). The energy balance equation is shown in Equation (2) and the corresponding bulk absorption coefficient, β, is presented in Equation (3) [77]:

$$Q_{IN} = Q_{ABS} + Q_{DEP} + Q_{REF} + Q_{LOST} \tag{2}$$

$$\beta_{ABS} + \beta_{DEP} + \beta_{REF} + \beta_{LOST} = 1 \tag{3}$$

where β_{ABS} refers to the bulk absorption coefficient due to Q_{ABS}, β_{DEP} refers to the bulk absorption coefficient due to Q_{DEP}, β_{REF} refers to the bulk absorption coefficient due to Q_{REF}, and β_{LOST} refers to the bulk absorption coefficient due to Q_{LOST}. From this study, it was experimentally proven that about

60% of the total initial energy was lost when using a laser as heat source, due to reflection by powders and also by powders not absorbing enough energy to melt.

2.5. Process–Microstructure Relationship

The energy source used during DED can be approximated as the Rosenthal solution of a moving heat source (laser, electron beam, plasma, or arc) on an infinite substrate. The microstructures obtained through DED can be predicted using the two important parameters: Thermal gradient, G (K/cm), and solidification front velocity (or interfacial velocity), R (cm/s). The relationship between G and R gives the thermal process maps [78,79]. Figure 4 shows graphs describing relationships between parameters, like G, R, arc length, undercooling, and supercooling. The mathematical relationships for the cooling rate, thermal gradient, and solidification front velocity are as follows (the reader could refer to [56] for a derivation of these equations):

$$\text{Cooling rate}: \frac{\partial T}{\partial t} \tag{4}$$

$$\text{Thermal Gradient}: G = |\nabla T| \tag{5}$$

$$\text{Solidification front velocity}: R = \frac{1}{G}\frac{\partial T}{\partial t} \tag{6}$$

where T is the temperature and t is the time. The melt pool circumference (MPC) length used in Figure 4 is the length of the circumference of the melt pool, and measurement starts from the bottom to the top of the melt pool in this study.

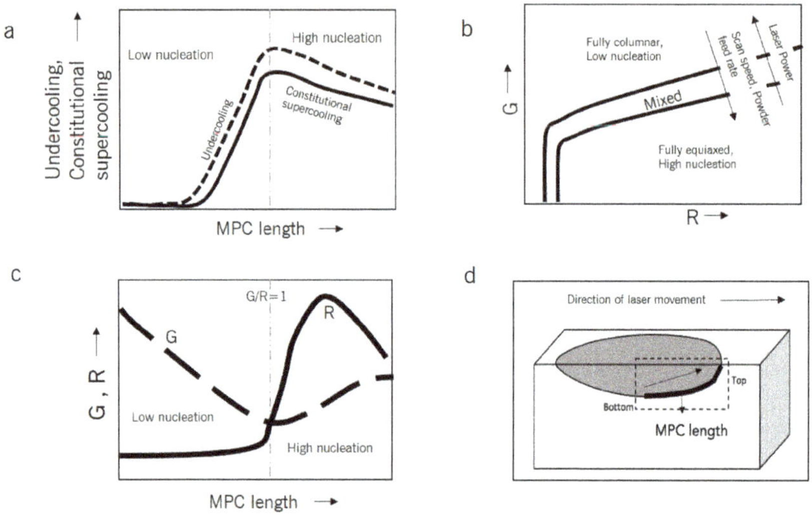

Figure 4. The relationships explaining the process maps for G vs. R, derived from [80]; (**a**) graph of undercooling/constitutional supercooling versus melt pool circumference (MPC) length showing the transition from low nucleation to high nucleation (columnar to equiaxed transition), (**b**) trend of the columnar to equiaxed transition from the graph of G versus R, (**c**) variation of the G/R ratio with respect to the MPC length, and (**d**) the MPC length as defined from the bottom to the top.

Some of the important trends derived from solidification front velocity and thermal gradient studies are:

- Cooling rates as high as 13,000 K/s were recorded in the literature during in-situ process monitoring of Ti-6Al-4V with laser based DED [81]. The initial cooling rates and microstructures changed with an increase in the number of deposited layers. This may be due to the accumulation of thermal energy in the part as more layers are deposited. This is proven in a study of Ti-6Al-4V printed with laser based DED, wherein a martensite microstructure formed in the first layers slowly convert to a Widmanstaten microstructure as the layers build up [82].
- According to the literature, the microstructure of laser based DED processed Ti-6Al-4V was columnar near the substrate (smaller MPC length) and equiaxed away from the substrate (higher MPC length) (due to differences in cooling rates), and there was a superheated melt pool during the process (almost 40%–50%) [81]. Figure 4a shows the relationship between undercooling and constitutional supercooling with respect to the MPC length.
- Heterogeneous nucleation varies with respect to the scanning speed, powder feed rate, and heat source power as shown in Figure 4b. Increases in the scanning speed and powder feed rate or decreasing the heat source power decreases the thermal gradient, G. That leads to an increase in R and a higher amount of equiaxed structure in the part.
- The ratio of G/R is an important parameter. $G/R > 1$ means a low rate of nucleation, due to a thermal gradient that is greater than the solidification front velocity, and $G/R < 1$ means a high rate of nucleation, due to a thermal gradient that is smaller than the solidification front velocity, as shown in Figure 4c.
- Figure 4d shows the MPC length, and increases in the value from the bottom to the top of the melt pool.

To conclude, as the MPC length increases, there is an increasing trend of undercooling and constitutional supercooling, leading to more columnar-to-equiaxed transition (CET), till the G/R ratio reaches 1. Therefore, CET is dictated by the thermal history and can be engineered according to requirements [80].

2.6. Dilution

DED involves the printing of subsequent layers of material to build a final part. There is heat accumulation due to multi-track and multi-layer deposition, and that will change the microstructure and final properties of the printed parts. A minimum level of metallurgical bonding is required between subsequent layers, which can be quantified using a dimensionless parameter called dilution. Dilution can be defined as:

$$D = \frac{d}{h+d} \tag{7}$$

where d is the depth of the melt pool below the substrate level, and h is the height of the material deposited above the substrate level. A high value of h correlates to a higher powder flow rate or lower energy input (corresponding to low dilution and lack of fusion between different layers), whereas a high value of d is due to a lower powder feed rate or higher energy input (corresponding to high dilution and keyholing phenomenon). There is an optimum value for dilution, which results in a good metallurgical bonding [83]. Figure 5b shows the optimal dilution level in a metal-substrate system generally between 10% and 30%, which is the standard value adopted by several researchers [84], and Figure 5a,c shows the extreme cases of dilution (<10% and >30%, respectively).

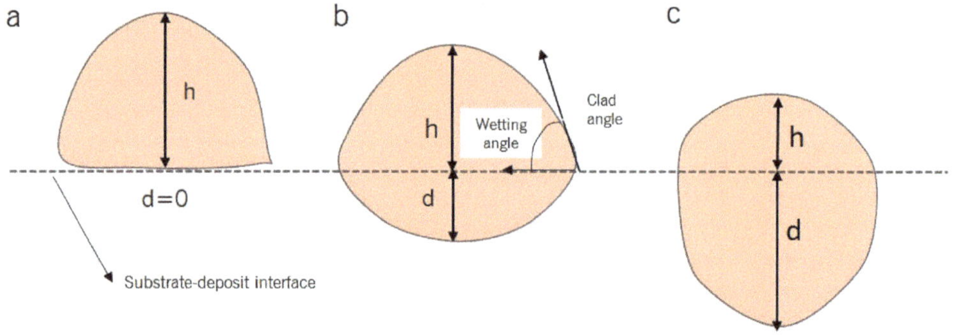

Figure 5. (**a**) Zero penetration ($d = 0$) or low penetration, meaning a lack of fusion with the substrate; (**b**) optimal level of dilution (generally between 10% and 30%); (**c**) high d leads to keyholing.

3. Common Defects

Defects may arise in the material during manufacturing. It is important to know the cause of such defects and take effective measures to counteract them. The most common defects arising during DED are discussed in the following sections.

3.1. Porosity

Porosity is one of the most important defects, which is linked to mechanical properties of DED processed parts. There are two major types of porosities: Interlayer porosity and intralayer porosity. Interlayer porosity occurs due to a lack of sufficient energy input to melt the filler material (powder or wire), leading to un-melted regions. This happens near the substrate or between un-melted tracks, when the linear heat input (heat source energy/scanning speed) is decreased or the mass flow rate is increased or a combination of the two. Interlayer pores are usually large and irregular in shape and occur due to higher solidification cooling rates. Low dilution values also cause the occurrence of interlayer porosities. Intralayer porosity is linked to the use of inert shielding gas during the DED process, promoting gas entrapment. Intralayer pores are usually spherical in shape and occur at random locations, owing to vaporization leading to gas trapped porosities, and observed within regions having lower solidification cooling rates [82,85,86]. High dilution refers to the occurrence of intralayer porosities. Interlayer and intralayer porosities are schematically shown in Figure 6a,b, respectively. Porosity is also dependent on the initial powder characteristics and uniformity [87]. If the starting powders do not have a uniform size distribution, it will give rise to more porosity in the final build. Inherent porosity inside the powder particle also leads to defects in the final part [88]. Taking all these modes of porosities into account, it is important to control the shape and size range of powders, and also maintain optimal process parameters during deposition. Porosity can be measured both qualitatively and quantitatively. Some commonly used techniques include the Archimedes principle, X-ray computed tomography, and optical microscopy.

Another important parameter in the literature, referred to as the global energy density (GED), establishes a relationship between interlayer (lack of fusion) and intralayer (keyholing) porosity:

$$\mathrm{GED} = \frac{P}{vd} \tag{8}$$

In Equation (8), GED is defined for laser based DED, where P is the laser power, v is the scan speed, and d is the laser spot size. GED can be easily correlated with dilution. As shown in Figure 6c, lower values of GED lead to less dilution (the negative slope), meaning more propensity to a lack of fusion defect, whereas higher values of GED lead to high dilution (the positive slope), meaning more tendency to form keyhole porosity [85].

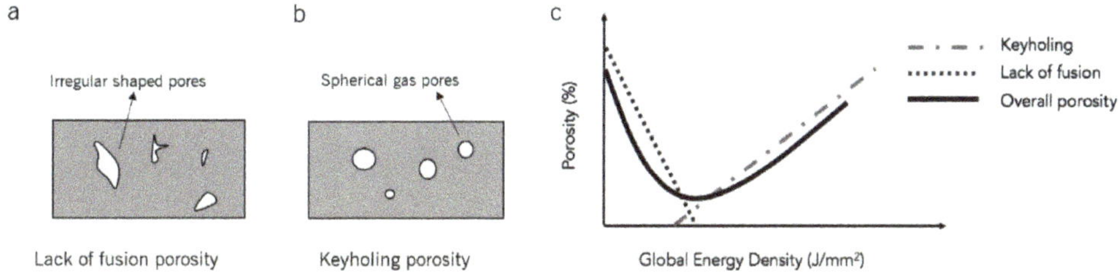

Figure 6. Schematic of: (**a**) Lack of fusion porosity (interlayer porosity), (**b**) keyholing porosity (intralayer porosity), and (**c**) the intersection of interlayer and intralayer porosity with respect to global energy density (GED).

3.2. Changes in Chemical Compositions due to Solute Segregation and Loss of Alloying Elements

When several layers are deposited during AM, it gives rise to the redistribution of solute particles, leading to the segregation and formation of heterogeneous microstructural bands, also known as heat affected zone (HAZ). This happens due to differences in the compositions at the interfaces of the substrate and deposit. The amount of solute segregation also depends upon the solidification cooling rates, with higher solidification cooling rates having a higher probability of a solute trap. These changes give rise to compositional inhomogenity along the printed material [89,90]. Another phenomenon, the loss of alloying elements, occurs due to the preferential vaporization of a few elements in the alloy during DED, due to differences in the boiling points of individual elements. For example, in a study of printed 304L stainless steels, compositional gradients developed along the build direction due to a greater loss of volatile elements (e.g., Cr, Mn, and Ni) as more heat built up in the system. The loss of these austenite stabilizers led to an increased hardness in the build direction due to the presence of a more martensitic phase in the upper layers [91]. The Langmuir equation can quantitatively predict the vaporization flux of the alloying elements, given by [65]:

$$J_i = \frac{\lambda P_i}{\sqrt{2\pi M_i T}} \quad (9)$$

where J_i is the vaporization flux of alloying elements, M_i is the molecular weight, P_i is the vapor pressure of the alloying elements, T is the temperature, and λ is a positive fraction, which estimates the condensation of some vaporized atoms. Consecutively, the mass of material vaporized can be estimated by:

$$\Delta m_i = \frac{L A_s J_i}{v} \quad (10)$$

where Δm_i is the mass vaporized, L is the track length, A_s is the melt pool area, v is the scan speed, and J_i is the vaporization flux of alloying elements. An example where the above equations were applied to estimate the amount of material lost during AM inferred that Al in Ti-6Al-4V is most susceptible to composition changes during DED, followed by Mn in stainless steel 316, with the least susceptible being Cr in Inconel 625 [92].

3.3. Printability of Alloys

Not all alloys are suitable to be processed by AM. Quantitatively, the printability of alloys could be defined using a dimensionless parameter known as thermal strain (Equation (11)). Lower values of thermal strain cause lesser residual stress in the material, thereby increasing the printability of the alloy by AM:

$$\varepsilon^* = \frac{\beta \Delta T}{EI} \frac{t}{F\sqrt{\rho}} H \quad (11)$$

where β is the bulk thermal expansion coefficient, ΔT is the change in temperature, t is the deposition time, H is the heat input, EI is the flexural rigidity, F is the Fourier number, and ρ is the density of material (the derivation of this particular equation has been performed in literature [92]). So, with an increase in $\beta\Delta T$, t, and H, there is an increase in the thermal strain, whereas an increase in EI and F can decrease thermal strain.

4. Mechanical Properties

4.1. Tensile Strength

The tensile strength and ductility of printed parts is dependent on the DED process parameters and the microstructure. There are several instances from the literature, which record varying trends of tensile behavior and ductility for the same material printed by DED. For instance, in one study, the tensile strength of DED fabricated Ti-6Al-4V was found to be similar to wrought manufactured Ti-6Al-4V, but with reduced ductility [93]. Another study showed that DED processed Ti-6Al-4V has a higher tensile strength due to a finer microstructure as compared with wrought alloy, but still exhibits lower ductility, due to a combination of the fine microstructure and the presence of internal defects [94]. Yet another study on DED processed parts showed an anisotropic porosity and tensile behavior in three different orientations, due to microstructural anisotropy [95]. The same study also demonstrated that with a 0.0124% increase in oxygen and decrease in the alpha lath width of DED processed Ti-6Al-4V, the yield strength and ultimate tensile strength (UTS) increased without any change in ductility. Post-processing, like heat treatments or hot isostatic pressing, tends to improve the ductility with a slight decrease in the tensile strength [94].

4.2. Hardness

Microhardness values can change along the build direction due to variations in the microstructure. The microhardness is higher near the first and the last layers, and relatively lower in the central layers. This variation could be attributed to the cyclic thermal history during the DED process. There is a higher heat buildup in the central layers, resulting in lower microhardness values [96]. A higher hardness and finer microstructure was reported in the literature by increasing the substrate thickness (more substrate mass acts as a faster heat sink). Increasing the substrate temperature decreased the hardness, due to a decrease in the thermal gradients and cooling rates, which can lead to a coarser microstructure. A higher interlayer dwell time is also known to increase the hardness, due to steeper thermal gradients [82]. Hardness studies by Zuback et al. stated that post processing of AM parts (like heat treatment or aging) or alloy selection could give more control over hardness, rather than changing the DED process parameters [97].

4.3. Fatigue

Fatigue is an important criterion for determining the structural integrity of materials printed by DED. The fatigue properties of AM parts have been studied by several groups [98–100] and recently reviewed by Bian et al. [101]. Fatigue is influenced significantly by the microstructure and defects [102]. In DED processed parts, the fatigue life can be estimated by determining the fatigue crack growth and number of probable fatigue initiation sites (or pores) [9]. Another fatigue initiator is un-melted powders, which could subsequently reduce the fatigue life by an order of magnitude [103]. In-situ high energy X-ray microtomography tests for fatigue crack propagation data at Argonne Photon Source (APS) were correlated with fatigue data from other conventional fatigue testing techniques (i.e., direct current potential drop techniques and fracture surface striations) [93]. It was found that fatigue crack growth was mostly in plane, with some cracks propagating towards the direction of tensile force. The crack growth rate was found to vary along different directions, and was also location dependent [93]. As of now, there is still a lack of consistency in the fatigue behavior reported by several authors. For instance, a study of LENS™ processed Ti-6Al-4V found a better high cycle fatigue life with respect to cast Ti-6Al-4V [104], while another study found a similar high cycle fatigue life for both LENS™ and wrought materials [103]. Another study stated that as-deposited DED shows similar properties to those of cast Ti-6Al-4V, while heat-treated DED Ti-6Al-4V has similar properties to those of wrought Ti-6Al-4V [94]. Hot isostatic pressing (HIP) improved the fatigue life of DED parts, by closing the porosities inside the parts [105]. As fatigue data is crucial to understanding the damage tolerance of structural materials, a lot more research is required in this field to establish accurate predictions of fatigue properties.

4.4. Residual Stress

Residual stress (RS) is generated during DED or any other metal AM technique due to the presence of steep thermal gradients between the heat source and the surrounding material. RS has the ability to damage the printed parts, due to distortion and cracking. RS is different at different locations in the printed metal. Studies have shown that residual stresses near the surface are tensile in nature, while the ones in the center are compressive stresses. RS tends to be higher when they occur between dissimilar materials [56]. Maximum RS was observed at the substrate–deposit interface. RS in metal AM can be broadly classified into two types based on the length scale: On the macroscale and on the microscale and nanoscale [106]. RS measurements on the macroscale are most widely used and can be conducted using non-destructive techniques, like X-ray diffraction and neutron diffraction [89]. Different methods can be used to relieve or reduce residual stresses. The most common techniques are: Preheating the substrate or preheating the initial feedstock to decrease the steep thermal gradients; using in situ process monitoring with feedback control to tune the process parameters on the fly; or using ex-situ post processing techniques, like heat treatment, to relief RS [106].

5. DED Process Control and Monitoring

Controlling the DED process is complicated due to the potential involvement of more than one type of material and also due to high build volumes, making it prone to defects. Also, the majority of users rely on expensive and time-consuming techniques, such as multiple experimental runs, to define optimized process parameters. The National Institute of Standards and Technology (NIST) highlights this issue and acts as a catalyst to resolve the issue of non-uniformity in printed parts by developing better process monitoring protocols for faster industrialization of DED [107]. For example, porosity control in a DED system is crucial, as it directly affects the structural integrity of the part. Some efforts made towards DED process control are tabulated in Table 4. However, these studies do not provide comprehensive information on how the material's thermo-mechanical properties change dynamically during the process. Hence, there is a need for the application of robust scientific techniques which could counter these limitations and help us monitor material related properties dynamically. Recent developments in quality monitoring includes high-energy X-ray synchrotron studies of DED. These encompass: High energy synchrotron X-ray source and high speed imaging camera used in tandem to detect the in situ melt pool geometries and deduce the phase transformations of Ti-6Al-4V [108]; a piezo driven powder delivery in conjunction with a laser heat source to investigate the powder–melt pool interaction during printing of Ti-6Al-4V [109]. These studies provide insights into the DED process physics, but are still far from mimicking all the components in a real DED system. Hence, there is still lack of data for real industrial DED systems and future research in this area is required.

Table 4. Selected studies on DED process monitoring.

Study Technique	Function	Reported Literature
In-situ acoustic monitoring	Detects defects and cracks during process	[110,111]
Non-destructive thermographic inspections like IR cameras	Captures the thermal history	[81,112]
IR two wave pyrometers and high-speed CCD cameras	Monitors the melt-pool characteristics and temperature	[81,113–118]
High energy X-ray diffraction and imaging	Detects phase transformations and melt pool dynamics	[73,108,109]
Repetitive process controller	Used to optimize layer height during the process	[119]

6. Determination of Optimal Process Parameters for Laser Based Powder-Fed DED

DED is an emerging field in the area of metal AM, and our goal was to create efficient process maps which provide a holistic picture of the DED process parameters. This is expected to save the user time, money, and effort to design their experiments. Inspiration was taken from the work of M. Thomas et al., who created normalized process diagrams for selective laser melting using dimensionless numbers [120]. However, DED has an additional parameter of the powder feed rate. After scrutiny, the linear heat input and powder feed rate were selected as primary parameters to make the process maps. The corresponding equations for these parameters are as follows:

$$\text{Linear heat input} = \frac{P}{v} \tag{12}$$

where P is the laser power, v is the scanning speed of the laser, and:

$$\text{Powder feed rate} = \frac{M}{t} \tag{13}$$

where M is the total mass of the powder and t is the time to deposit powders. The linear heat input is a standard parameter which has been used often in the literature, normalized using the ratio of the two fundamental parameters: Laser power and scanning speed [74]. The linear heat input can be used for any range of values for the power and scanning speed, and it has been proven experimentally that the same linear power density gives similar properties [121]. Some previous studies have attempted to build DED process maps, based on the linear mass density [122,123], where:

$$\text{Linear Mass Density} = \frac{\frac{M}{t}}{v} \tag{14}$$

However, the linear mass density is not a robust parameter, as it fails to consider the influence of the effective residence time of the laser beam spot per unit volume of the powder feed. A higher scan speed or a higher powder feed rate will result in a lower effective residence time of the laser spot per unit volume of powder. To understand the consequences due to a less effective residence time, consider two extreme cases of process parameters, one having a high speed and high mass flow rate, and the other having a low speed and low mass flow rate. Even though both cases produce the same value of linear mass density, their properties are significantly different [122]. The high speed and high mass flow rate case will have worse properties due to the much lower effective residence time. To counteract such problems associated with the linear mass density and to be consistent irrespective of the varying range of values, the linear heat input was chosen to construct process maps.

The other unique DED parameter is the powder feed rate, which is not applicable for other AM systems, like powder bed fusion, selective laser melting, etc. Powder feed rate values will determine how much powder is transferred to the laser spot area. The powder catchment efficiency varies, depending upon process parameters, like the feed rate, carrier gas flow rate, particle size, and velocity, etc., and it determines the percentage of the total powders that gets inside the melt-pool [59]. A material utilization efficiency of about 70% to 90% was reported for DED of Ti-6Al-4V powders [124]. Insufficient heat input or a very high powder feed rate also leads to unmelted powders.

A graph of the linear heat input versus the powder feed rate is plotted using selected data points from the literature, as listed in Table 5. The criterion for the selection of these specific data points was on the basis of their corresponding values of dilution. As mentioned in Section 2.5, about a 10% to 30% dilution represents a good amount of metallurgical bonding between subsequent layers (the clad dimensions are related to the scanning speed and powder feed rate of the DED process. These consecutively affect the contact angle, which determines the bonding of the deposit to the substrate, and overall, these quantities can be understood through the dilution parameter). However, there might be exceptions to the optimum dilution range. For example, Ti-15Mo alloy requires higher values of dilution to achieve optimal conditions due to the refractory nature of the material. Wherever no information was given about the optimal dilution values, it was considered to have an optimal dilution of 10% to 30%. Qualitatively, if the dilution level goes above 30%, it might lead to keyholing or below 10% might lead to a lack of fusion, and both cases are considered to be outside the optimal processing range. Another region on the process map is identified, called the mixed-mode porosity region, that occurs due to the combination of both keyholing and a lack of fusion. The high linear heat input is excessive for the upper layers of the powders, but due to the high feed rate, the heat input does not melt the bottom layers very efficiently (shielding by the powders). As a result, the top powder volume experiences keyholing and the bottom powder volume experiences a lack of fusion. A gas tungsten arc welding study defined this mid-porosity region in the literature [125]. This unique resemblance can be attributed towards the similarities between the welding and DED processes.

Table 5. Compilation of optimal processing data point ranges for DED deposited metal or alloy systems.

Material System	Optimal Dilution Range or Optimal Process Parameters or Acceptable Range Values as Reported	Study
Stainless steel grade 303L	–	[126]
H13 tool steel	Optimal, generally low dilution preferred	[12]
Inconel 718 in steel substrate	–	[127]
Ti-6Al-4V	Within 10%–30%	[128]
Ti-6Al-4V	–	[5]
H13 tool steel	–	[129]
P420 steel on low carbon steel	12%–20%; optimal is 16%	[130]
Inconel 690 on Inconel 600	Optimal	[131]
24CrMoNi alloy steel	32%	[2]
Ti-15Mo	54%–68%	[132]
Fe	10%–20%	[133]
Inconel 718 on AISI 1045 carbon steel	–	[83]
Stellite® 12 (Cr-W based alloy) on 0.2% C steel	–	[134]
Zirconium on Zr alloy	50%	[135]
Ti-6Al-4V	48%–52%	[136]
NiFeBSiNb on mild steel	Above 10%	[137]
Inconel	Optimal	[121]
Fe	–	[138]
Ti-6Al-4V	Within 10% to 30%	[128]
Ti-6Al-4V	Zero porosity data; optimal	[139]
NiCrAlY on Inconel 738 superalloy	15%–25%	[140]
Ni-Cr based alloy (commercially known as 19E alloy)	5%–30%	[141]

The graph shown in Figure 7 gives the optimal processing ranges for the alloy systems, like Inconel, Ti-6Al-4V, H13 Tool steel, Fe, Ti-15Mo, and some Ni-Cr alloys. From the graph, three regions which do not contain any optimal data points are observed. It can be hypothesized that these regions are due to three modes of increased porosity formation in the material: Keyholing (due to a high linear heat input and low powder feed rate), lack of fusion (due to a low linear heat input and high powder feed rate), and mid porosity zone (due to an appreciably high linear heat input and high powder feed rate contributing to mixed-mode porosity). These regions have been defined up to a linear heat input of 400 J/mm and a powder feed rate up to 25 g/min. Such regions are valid for most of the metal-alloy systems that have been studied.

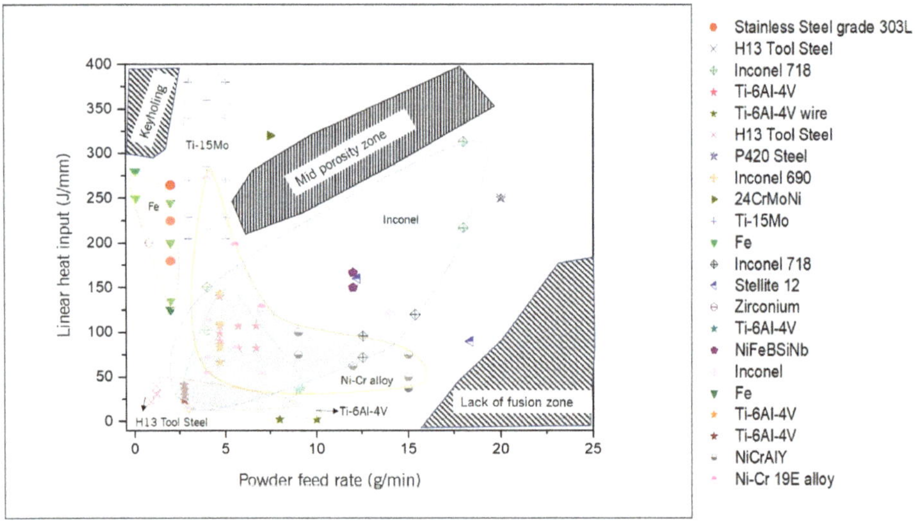

Figure 7. Optimal processing regions for Ti-6Al-4V, Inconel 690-718, Ni-Cr based alloys, Fe, H13 tool steel, and Ti-15Mo alloy, along with the unsuitable processing zones, as compiled in Table 5.

7. Applications and Emerging Technologies

This paper discussed the physics of DED technology and established process maps, which will be useful in various research fields as well as industries. This section lists some applications, both common in other metal AM techniques, as well as ones which are unique to DED, aiding the readers to understand the diverse functionalities of DED.

7.1. DED Metal Parts Used in Various Biomedical Applications

It is advantageous to produce porous implants using DED as compared to conventional casting methods: It is possible to alter the mechanical properties simply by changing the orientation or geometry of the build; it is possible to incorporate different materials together and obtain the optimal properties through a functionally graded material; and it is much easier to custom build the implants due to specific patient requirements. The most common materials used for biomedical applications are Ti and its alloys, Co based alloys, 316L stainless steel, and Ni-Ti based alloys. Additive manufactured parts have gained prominence in the orthopedic and dental implant industry. Biomedical applications garnered a revenue of 16.4% of the total AM industry in 2012 [10]. This shows promising metal DED applications in the biomedical industry, with a growing trend in the future, specifically in dental [142], orthopedic, and cardiovascular [143] applications. Biocompatibility tests on porous Ti-6Al-4V made with LENS™ proved the capability of cell growth on implants having a pore size of 200 μm or larger [144]. Also, in-vivo biocompatibility studies with porous Ti-6Al-4V processed by DED showed that a pore volume fraction of 0.40 (upper limit) can accelerate the healing process through biological fixation [144].

7.2. Welding and Cladding

Conventional welding leads to high RS at the interfaces, especially for the welding of dissimilar metals. This might lead to early failures, and the results could be catastrophic. DED can be used to change the composition as a function of the position, facilitating a smooth transition from one joint to the other. This could be achieved by designing a gradient path that avoids the unwanted phases determined from multi-component phase diagrams [145]. This will reduce the RS and improve the mechanical integrity of the joints [146–152]. Cladding is generally used to form corrosion resistant protective coatings on substrates or to improve the tribological properties [56,153,154]. There is also an added advantage of using DED for cladding two dissimilar materials, due to the possibility of using functionally graded alloys. Another useful cladding technique is multi-axis cladding, making it possible to deposit layers at any angular axis. This functionality is a great advantage of DED over other AM systems [56,155–157].

7.3. Repair

Repairing (or remanufacturing) is essential to improve the life cycle of parts and to restore their functionality. This also leads to a reduced environmental impact, due to less material and energy wastage [12,13]. DED is a well-known repair technique, with the parts possessing good metallurgical bonding and exhibiting good post-repair mechanical properties [4,158]. Some studies establishing DED as a standard repairing technique in industries include: Repairing a gas-turbine blade using a Ni based superalloy delivered through a co-axial powder feeder [159]; repairing steam circuit parts at thermal power stations, using deposition of Co based alloys to maintain high temperature mechanical properties [160]; and repairing of Ti-6Al-4V aero engine parts using Ti-6Al-4V powders [161].

7.4. Bulk Combinatorial Alloy Design

It is possible to design alloys with compositional gradients using DED and this functionality is a unique characteristic which distinguishes DED from other AM systems. For example, the Ti-6Al-4V to V gradient and 304L steel to Invar 36 gradient were processed in the literature using DED. The aim was to successfully design the gradient path such that the unwanted brittle phases could be avoided in the microstructure, which would eventually give better mechanical properties for the printed parts [145]. This was done using multi-component phase diagrams. Another study used Cr-V-Mo hot working tool steel and Ni based maraging steel as base materials, with varying ratios of these two materials [162]. It was hot rolled and subsequently characterized, enabling high throughput probing of important alloy blends. DED processed high entropy alloy (HEA) AlCrFeMoVx ($x = 0$ to 1) was also studied in literature, to assess the composition–microstructure–hardness relationships [163]. High hardness was observed with increasing V, due to the high V solubility in this HEA leading to solid solution strengthening.

7.5. Construction Materials

AM, and in general DED, has good potential in the construction industry, but has its fair share of challenges, as structural members are usually quite big to be built using AM. Nonetheless, it would be advantageous to build highly specialized parts by exploiting this technology. Conventional casting leads to prismatic structures (uniform microstructures). With DED, engineered compositional and microstructural gradients in the structural parts are possible, which might give superior mechanical properties. An important thing to note is that the construction industry contributes to about 30% of greenhouse emissions in total, which could be reduced by partly adopting AM for the mass production of specialized parts [164].

7.6. Hybrid Additive Manufacturing

DED printed parts have several problems associated with it, like RS, lack of surface finish, etc., and to address these challenges without separate post-processing, several hybrid AM systems have been developed. Hybrid CNC-AM systems fully integrate the capabilities of both additive and subtractive manufacturing, which can be further exploited to increase productivity and competitiveness in the market. The parts produced by this hybrid method are precise even when produced in large-scale, due to the post-processing techniques integrated into a single system, without the need for separate machining. Hybrid manufacturing is still a relatively new technology and requires a lot more research for acceptance into the market. Therefore, the most important steps to be taken in the direction of improvement of such machines would be process optimization strategies along with developments in software integration [165]. Other secondary processes that complement the performance of DED parts (to name a few) are: Inter-pass rolling [166] and Ultrasonic Vibration Assisted LENS™ [167], both used for grain size refinement to enhance the mechanical properties of the parts.

8. Summary and Outlook

There are many far-from-equilibrium and highly dynamic phenomena during DED due to extreme heating and cooling rates. These include dynamic melt pool, melting and vaporization of powder particles, rapid solidification, and phase transformation. Such transient events often result in a large scatter in mechanical properties of printed components due to many complex interactions, leading to unwanted phase transformations and grain structures, residual stresses, and porosities. Further studies on establishing a correlation between composition, process parameters (powder feed rate, laser power, and velocity), process signature (melt pool stability and dimensions), and the resultant microstructure, pore content, residual stresses, and macroscopic properties will be extremely beneficial to the advancement of this technology. It is expected that the process physics of alloy systems and composite systems would be different and future studies are required in each area both experimentally and computationally.

The major contribution in this paper was the establishment of process maps for DED, after compiling the available literature. Researchers will be able to use this map to predict their preferred operating ranges for different alloy classes, but further work is required to extend our study to more extensive material systems. Hybrid AM technologies were discussed toward the end of this paper. These are relatively new approaches to overcome some of the limitations of AM. Further research is required in this area to mature these technologies beyond the current state of the art.

Funding: This research received no external funding.

Conflicts of Interest: The authors declare no conflict of interest.

References

1. Mazumder, J. Design for metallic additive manufacturing machine with capability for "certify as you build". *Procedia CIRP* **2015**, *36*, 187–192. [CrossRef]
2. Cao, L.; Chen, S.; Wei, M.; Guo, Q.; Liang, J.; Liu, C.; Wang, M. Effect of laser energy density on defects behavior of direct laser depositing 24CrNiMo alloy steel. *Opt. Laser Technol.* **2019**, *111*, 541–553. [CrossRef]
3. Khairallah, S.A.; Anderson, A.T.; Rubenchik, A.M.; King, W.E. Laser powder-bed fusion additive manufacturing: Physics of complex melt flow and formation mechanisms of pores, spatter, and denudation zones. *Acta Mater.* **2016**, *108*, 36–45. [CrossRef]
4. Thompson, S.M.; Bian, L.; Shamsaei, N.; Yadollahi, A. An overview of Direct Laser Deposition for additive manufacturing; Part I: Transport phenomena, modeling and diagnostics. *Addit. Manuf.* **2015**, *8*, 36–62. [CrossRef]
5. Caiazzo, F. Additive manufacturing by means of laser-aided directed metal deposition of titanium wire. *Int. J. Adv. Manuf. Technol.* **2018**, *96*, 2699–2707. [CrossRef]

6. DebRoy, T.; Wei, H.L.; Zuback, J.S.; Mukherjee, T.; Elmer, J.W.; Milewski, J.O.; Beese, A.M.; Wilson-Heid, A.; De, A.; Zhang, W. Additive manufacturing of metallic components—Process, structure and properties. *Prog. Mater. Sci.* **2018**, *92*, 112–224. [CrossRef]
7. Sames, W.J.; List, F.A.; Pannala, S.; Dehoff, R.R.; Babu, S.S. The Metallurgy and Processing Science of metal additive manufacturing. *Int. Mater. Rev.* **2016**, *61*, 315–360. [CrossRef]
8. Gorsse, S.; Hutchinson, C.; Gouné, M.; Banerjee, R. Additive manufacturing of metals: A brief review of the characteristic microstructures and properties of steels, Ti-6Al-4V and high-entropy alloys. *Sci. Technol. Adv. Mater.* **2017**, *18*, 584–610. [CrossRef]
9. Lewandowski, J.J.; Seifi, M. Metal Additive Manufacturing: A Review of Mechanical Properties. *Annu. Rev. Mater. Res.* **2016**, *46*, 151–186. [CrossRef]
10. Harun, W.S.W.; Kamariah, M.S.I.N.; Muhamad, N.; Ghani, S.A.C.; Ahmad, F.; Mohamed, Z. A review of powder additive manufacturing processes for metallic biomaterials. *Powder Technol.* **2018**, *327*, 128–151. [CrossRef]
11. Pinkerton, A.J.; Ul Haq Syed, W.; Li, L. An experimental and theoretical investigation of combined gas- and water-atomized powder deposition with a diode laser. *J. Laser Appl.* **2006**, *18*, 73–80. [CrossRef]
12. Shin, Y.C.; Bailey, N.; Katinas, C.; Tan, W. Predictive modeling capabilities from incident powder and laser to mechanical properties for laser directed energy deposition. *Comput. Mech.* **2018**, *61*, 617–636. [CrossRef]
13. Ramani, K.; Ramanujan, D.; Bernstein, W.Z.; Zhao, F.; Sutherland, J.; Handwerker, C.; Choi, J.-K.; Kim, H.; Thurston, D. Integrated Sustainable Life Cycle Design: A Review. *J. Mech. Des.* **2010**, *132*, 091004. [CrossRef]
14. Greer, C.; Nycz, A.; Noakes, M.; Richardson, B.; Post, B.; Kurfess, T.; Love, L. Introduction to the design rules for Metal Big Area Additive Manufacturing. *Addit. Manuf.* **2019**, *27*, 159–166. [CrossRef]
15. Heralić, A.; Christiansson, A.K.; Lennartson, B. Height control of laser metal-wire deposition based on iterative learning control and 3D scanning. *Opt. Lasers Eng.* **2012**, *50*, 1230–1241. [CrossRef]
16. Zhang, L.C.; Liu, Y.; Li, S.; Hao, Y. Additive Manufacturing of Titanium Alloys by Electron Beam Melting: A Review. *Adv. Eng. Mater.* **2018**, *20*, 1–16. [CrossRef]
17. Hoefer, K. Arc based additive manufacturing of steel components—Comparison of wire- and powder based variants. *Weld. World* **2018**, *62*, 243–247. [CrossRef]
18. Moridi, A.; Hassani-Gangaraj, S.M.; Guagliano, M.; Dao, M. Cold spray coating: Review of material systems and future perspectives. *Surf. Eng.* **2014**, *30*, 369–395. [CrossRef]
19. Hassani-Gangaraj, M.; Veysset, D.; Champagne, V.K.; Nelson, K.A.; Schuh, C.A. Adiabatic shear instability is not necessary for adhesion in cold spray. *Acta Mater.* **2018**, *158*, 430–439. [CrossRef]
20. Zheng, B.; Topping, T.; Smugeresky, J.E.; Zhou, Y.; Biswas, A.; Baker, D.; Lavernia, E.J. The influence of Ni-coated TiC on laser-deposited IN625 metal matrix composites. *Metall. Mater. Trans. A* **2010**, *41*, 568–573. [CrossRef]
21. Li, X.C.; Stampfl, J.; Prinz, F.B. Mechanical and thermal expansion behavior of laser deposited metal matrix composites of Invar and TiC. *Mater. Sci. Eng. A* **2002**, *282*, 86–90. [CrossRef]
22. Liu, W.; DuPont, J.N. Fabrication of carbide-particle-reinforced titanium aluminide-matrix composites by laser-engineered net shaping. *Metall. Mater. Trans. A* **2004**, *35*, 1133–1140. [CrossRef]
23. Liu, W.; DuPont, J.N. Fabrication of functionally graded TiC/Ti composites by Laser Engineered Net Shaping. *Scr. Mater.* **2003**, *48*, 1337–1342. [CrossRef]
24. Gualtieri, T.; Bandyopadhyay, A. Additive manufacturing of compositionally gradient metal-ceramic structures: Stainless steel to vanadium carbide. *Mater. Des.* **2018**, *139*, 419–428. [CrossRef]
25. Zhang, Y.; Bandyopadhyay, A. Direct fabrication of compositionally graded Ti-Al2O3 multi-material structures using Laser Engineered Net Shaping. *Addit. Manuf.* **2018**, *21*, 104–111. [CrossRef]
26. Balla, V.K.; DeVasConCellos, P.D.; Xue, W.; Bose, S.; Bandyopadhyay, A. Fabrication of compositionally and structurally graded Ti-TiO2 structures using laser engineered net shaping (LENS). *Acta Biomater.* **2009**, *5*, 1831–1837. [CrossRef] [PubMed]
27. Bandyopadhyay, P.P.; Balla, V.K.; Bose, S.; Bandyopadhyay, A. Compositionally graded aluminum oxide coatings on stainless steel using laser processing. *J. Am. Ceram. Soc.* **2007**, *90*, 1989–1991. [CrossRef]
28. Das, M.; Balla, V.K.; Kumar, T.S.S.; Manna, I. Fabrication of Biomedical Implants using Laser Engineered Net Shaping (LENS™). *Trans. Indian Ceram. Soc.* **2013**, *72*, 169–174. [CrossRef]
29. Heer, B.; Bandyopadhyay, A. Silica coated titanium using Laser Engineered Net Shaping for enhanced wear resistance. *Addit. Manuf.* **2018**, *23*, 303–311. [CrossRef]

30. Das, M.; Balla, V.K.; Kumar, T.S.S.; Bandyopadhyay, A.; Manna, I. Tribological, electrochemical and in vitro biocompatibility properties of SiC reinforced composite coatings. *Mater. Des.* **2016**, *95*, 510–517. [CrossRef]
31. Bandyopadhyay, A.; Dittrick, S.; Gualtieri, T.; Wu, J.; Bose, S. Calcium phosphate-titanium composites for articulating surfaces of load-bearing implants. *J. Mech. Behav. Biomed. Mater.* **2016**, *57*, 280–288. [CrossRef] [PubMed]
32. Stenberg, K.; Dittrick, S.; Bose, S.; Bandyopadhyay, A. Influence of simultaneous addition of carbon nanotubes and calcium phosphate on wear resistance of 3D-printed Ti6Al4V. *J. Mater. Res.* **2018**, *33*, 2077–2086. [CrossRef]
33. Traxel, K.D.; Bandyopadhyay, A. Reactive-deposition based additive manufacturing of Ti-Zr-BN composites. *Addit. Manuf.* **2018**, *24*, 353–363. [CrossRef]
34. Ke, D.; Vu, A.A.; Bandyopadhyay, A.; Bose, S. Compositionally graded doped hydroxyapatite coating on titanium using laser and plasma spray deposition for bone implants. *Acta Biomater.* **2019**, *84*, 414–423. [CrossRef] [PubMed]
35. Torgerson, T.B.; Mantri, S.A.; Banerjee, R.; Scharf, T.W. Room and elevated temperature sliding wear behavior and mechanisms of additively manufactured novel precipitation strengthened metallic composites. *Wear* **2019**, *426*, 942–951. [CrossRef]
36. Srinivas, V.; Savitha, U.; Jagan Reddy, G. Processing and Characterization of NiCr-YSZ Compositionally Graded Coatings on Superalloy using Laser Engineered Net Shaping (LENS). *Mater. Today Proc.* **2018**, *5*, 27277–27284. [CrossRef]
37. Hu, Y.; Wang, H.; Li, Y.; Ning, F.; Cong, W. Surface grinding of ZTA parts fabricated by laser engineered net shaping process: Effects of ZrO2 content and ultrasonic vibration. In Proceedings of the 13th International Manufacturing Science and Engineering Conference, College Station, TX, USA, 18–22 June 2018.
38. Yan, J.; Masoudi, N.; Battiato, I.; Fadel, G. Optimization of process parameters in laser engineered Net shaping (LENS) deposition of multi-materials. In Proceedings of the ASME 2015 International Design Engineering Technical Conferences and Computers and Information in Engineering Conference, Boston, MA, USA, 2–5 August 2015; p. V01AT02A034.
39. Zhang, Y.; Sahasrabudhe, H.; Bandyopadhyay, A. Additive manufacturing of Ti-Si-N ceramic coatings on titanium. *Appl. Surf. Sci.* **2015**, *346*, 428–437. [CrossRef]
40. Bernard, S.A.; Balla, V.K.; Bose, S.; Bandyopadhyay, A. Direct laser processing of bulk lead zirconate titanate ceramics. *Mater. Sci. Eng. B* **2010**, *172*, 85–88. [CrossRef]
41. Li, Y.; Hu, Y.; Cong, W.; Zhi, L.; Guo, Z. Additive manufacturing of alumina using laser engineered net shaping: Effects of deposition variables. *Ceram. Int.* **2017**, *43*, 7768–7775. [CrossRef]
42. Niu, F.; Wu, D.; Zhou, S.; Ma, G. Power prediction for laser engineered net shaping of Al_2O_3 ceramic parts. *J. Eur. Ceram. Soc.* **2014**, *34*, 3811–3817. [CrossRef]
43. Wu, D.J.; Niu, F.Y.; Ma, G.Y.; Zhang, B.; Yan, S. Process optimization for suppressing cracks in laser engineered net shaping of Al_2O_3 ceramics. *JOM* **2016**, *69*, 557–562.
44. Ma, G.; Wang, J.; Niu, F.; Sun, B.; Wu, D. Influence of powder distribution on the Al_2O_3 thin-wall ceramic formed by laser engineered net shaping. *Chin. J. Lasers* **2015**, *42*, 0103006.
45. Niu, F.; Wu, D.; Ma, G.; Zhang, B. Additive manufacturing of ceramic structures by laser engineered net shaping. *Chin. J. Mech. Eng.* **2015**, *28*, 1117–1122. [CrossRef]
46. Wu, D.; Liu, H.; Lu, F.; Ma, G.; Yan, S.; Niu, F.; Guo, D. Al_2O_3-YAG eutectic ceramic prepared by laser additive manufacturing with water-cooled substrate. *Ceram. Int.* **2019**, *45*, 4119–4122. [CrossRef]
47. Roy, M.; Vamsi Krishna, B.; Bandyopadhyay, A.; Bose, S. Laser processing of bioactive tricalcium phosphate coating on titanium for load-bearing implants. *Acta Biomater.* **2008**, *4*, 324–333. [CrossRef] [PubMed]
48. Yan, S.; Wu, D.; Niu, F.; Huang, Y.; Liu, N.; Ma, G. Effect of ultrasonic power on forming quality of nano-sized Al_2O_3-ZrO_2 eutectic ceramic via laser engineered net shaping (LENS). *Ceram. Int.* **2018**, *44*, 1120–1126. [CrossRef]
49. Hu, Y.; Ning, F.; Cong, W.; Li, Y.; Wang, X.; Wang, H. Ultrasonic vibration-assisted laser engineering net shaping of ZrO_2-Al_2O_3 bulk parts: Effects on crack suppression, microstructure, and mechanical properties. *Ceram. Int.* **2018**, *44*, 2752–2760. [CrossRef]
50. Niu, F.; Wu, D.; Ma, G.; Wang, J.; Guo, M.; Zhang, B. Nanosized microstructure of Al_2O_3-ZrO_2 (Y_2O_3) eutectics fabricated by laser engineered net shaping. *Scr. Mater.* **2015**, *95*, 39–41. [CrossRef]
51. Marattukalam, J.J.; Singh, A.K.; Datta, S.; Das, M.; Balla, V.K.; Bontha, S.; Kalpathy, S.K. Microstructure and corrosion behavior of laser processed NiTi alloy. *Mater. Sci. Eng. C* **2015**, *57*, 309–313. [CrossRef]

52. Stull, J.A.; Hill, M.A.; Lienert, T.J.; Tokash, J.; Bohn, K.R.; Hooks, D.E. Corrosion characteristics of laser-engineered net shaping additively-manufactured 316L stainless steel. *JOM* **2018**, *70*, 2677–2683. [CrossRef]
53. Dehoff, R.R.; Sarosi, P.M.; Collins, P.C.; Fraser, H.L.; Mills, M.J. Microstructural evaluation of LENS™ deposited Nb-Ti-Si-Cr alloys. *MRS Online Proc. Libr. Arch.* **2002**, *753*, 2–7. [CrossRef]
54. Sridharan, N.; Cakmak, E.; Dehoff, R.R. Microstructure evolution during laser direct energy deposition of a novel Fe-Cr-Ni-W-B hardfacing coating. *Surf. Coat. Technol.* **2019**, *358*, 362–370. [CrossRef]
55. Gu, D.D.; Meiners, W.; Wissenbach, K.; Poprawe, R. Laser additive manufacturing of metallic components: Materials, processes and mechanisms. *Int. Mater. Rev.* **2012**, *57*, 133–164. [CrossRef]
56. Gibson, I.; Rosen, D.; Stucker, B. *Additive Manufacturing Technologies: 3D Printing, Rapid Prototyping, and Direct Digital Manufacturing*, 2nd ed.; Springer: New York, NY, USA, 2015.
57. DebRoy, T.; Zhang, W.; Turner, J.; Babu, S.S. Building digital twins of 3D printing machines. *Scr. Mater.* **2017**, *135*, 119–124. [CrossRef]
58. Pinkerton, A.J.; Moat, R.; Shah, K.; Li, L.; Preuss, M.; Withers, P.J. A verified model of laser direct metal deposition using an analytical enthalpy balance method. In Proceedings of the International Congress on Applications of Lasers & Electro-Optics, Orlando, FL, USA, 29 October–1 November 2007.
59. Liu, S.; Zhang, Y.; Kovacevic, R. Numerical simulation and experimental study of powder flow distribution in high power direct diode laser cladding process. *Lasers Manuf. Mater. Process.* **2015**, *2*, 199–218. [CrossRef]
60. Heigel, J.C.; Michaleris, P.; Reutzel, E.W. Thermo-mechanical model development and validation of directed energy deposition additive manufacturing of Ti-6Al-4V. *Addit. Manuf.* **2015**, *5*, 9–19. [CrossRef]
61. Kovaleva, I.; Kovalev, O.; Zaitsev, A.; Smurov, I. Numerical simulation and comparison of powder jet profiles for different types of coaxial nozzles in direct material deposition. *Phys. Procedia* **2013**, *41*, 870–872. [CrossRef]
62. Tian, Y.; McAllister, D.; Colijn, H.; Mills, M.; Farson, D.; Nordin, M.; Babu, S. Rationalization of microstructure heterogeneity in INCONEL 718 builds made by the direct laser additive manufacturing process. *Metall. Mater. Trans. A Phys. Metall. Mater. Sci.* **2014**, *45*, 4470–4483. [CrossRef]
63. Radhakrishnan, B.; Gorti, S.; Babu, S. Large scale phase field simulations of microstructure evolution during thermal cycling. In Proceedings of the International Conference on Solid-Solid Phase Transformations in Inorganic Materials (PTM), Whistler, BC, Canada, 28 June–3 July 2015.
64. Amine, T.; Newkirk, J.W.; Liou, F. Investigation of effect of process parameters on multilayer builds by direct metal deposition. *Appl. Therm. Eng.* **2014**, *73*, 500–511. [CrossRef]
65. DebRoy, T.; David, S.A. Physical processes in fusion welding. *Rev. Mod. Phys.* **1995**, *67*, 85–112. [CrossRef]
66. Raghavan, A.; Wei, H.L.; Palmer, T.A.; DebRoy, T. Heat transfer and fluid flow in additive manufacturing. *J. Laser Appl.* **2014**, *25*, 052006. [CrossRef]
67. Manvatkar, V.; De, A.; DebRoy, T. Spatial variation of melt pool geometry, peak temperature and solidification parameters during laser assisted additive manufacturing process. *Mater. Sci. Technol.* **2015**, *31*, 924–930. [CrossRef]
68. Manvatkar, V.; De, A.; Debroy, T. Heat transfer and material flow during laser assisted multi-layer additive manufacturing. *J. Appl. Phys.* **2014**, *116*, 124905. [CrossRef]
69. Farahmand, P.; Kovacevic, R. An experimental–numerical investigation of heat distribution and stress field in single- and multi-track laser cladding by a high-power direct diode laser. *Opt. Laser Technol.* **2014**, *63*, 154–168. [CrossRef]
70. Cao, J.; Gharghouri, M.A.; Nash, P. Finite-element analysis and experimental validation of thermal residual stress and distortion in electron beam additive manufactured Ti-6Al-4V build plates. *J. Mater. Process. Technol.* **2016**, *237*, 409–419. [CrossRef]
71. Heiple, C.R.; Roper, J.R.; Stagner, R.T.; Aden, R.J. Surface active element effects on the shape of GTA, laser and electron beam welds. *Weld. J.* **1983**, *62*, 72–77.
72. Mills, K.C.; Keene, B.J.; Brooks, R.F.; Shirali, A. Marangoni effects in welding. *Philos. Trans. R. Soc. A Math. Phys. Eng. Sci.* **1998**, *356*, 911–925. [CrossRef]
73. Aucott, L.; Dong, H.; Mirihanage, W.; Atwood, R.; Kidess, A.; Gao, S.; Wen, S.; Marsden, J.; Feng, S.; Tong, M.; et al. Revealing internal flow behaviour in arc welding and additive manufacturing of metals. *Nat. Commun.* **2018**, *9*, 1–7. [CrossRef]
74. Mukherjee, T.; Manvatkar, V.; De, A.; DebRoy, T. Dimensionless numbers in additive manufacturing. *J. Appl. Phys.* **2017**, *121*, 024904. [CrossRef]

75. Van Elsen, M.; Al-Bender, F.; Kruth, J.P. Application of dimensional analysis to selective laser melting. *Rapid Prototyp. J.* **2008**, *14*, 15–22. [CrossRef]
76. Mukherjee, T.; Manvatkar, V.; De, A.; DebRoy, T. Mitigation of thermal distortion during additive manufacturing. *Scr. Mater.* **2017**, *127*, 79–83. [CrossRef]
77. Lia, F.; Park, J.; Tressler, J.; Martukanitz, R. Partitioning of laser energy during directed energy deposition. *Addit. Manuf.* **2017**, *18*, 31–39. [CrossRef]
78. Song, J.; Chew, Y.; Bi, G.; Yao, X.; Zhang, B.; Bai, J.; Moon, S.K. Numerical and experimental study of laser aided additive manufacturing for melt-pool profile and grain orientation analysis. *Mater. Des.* **2018**, *137*, 286–297. [CrossRef]
79. Bontha, S.; Klingbeil, N.W.; Kobryn, P.A.; Fraser, H.L. Thermal process maps for predicting solidification microstructure in laser fabrication of thin-wall structures. *J. Mater. Process. Technol.* **2006**, *178*, 135–142. [CrossRef]
80. Liu, P.; Wang, Z.; Xiao, Y.; Horstemeyer, M.F.; Cui, X.; Chen, L. Insight into the mechanisms of columnar to equiaxed grain transition during metallic additive manufacturing. *Addit. Manuf.* **2019**, *26*, 22–29. [CrossRef]
81. Marshall, G.J.; Young, W.J.; Thompson, S.M.; Shamsaei, N.; Daniewicz, S.R.; Shao, S. Understanding the microstructure formation of Ti-6Al-4V during direct laser deposition via in-situ thermal monitoring. *JOM* **2016**, *68*, 778–790. [CrossRef]
82. Kistler, N.A.; Corbin, D.J.; Nassar, A.R.; Reutzel, E.W.; Beese, A.M. Effect of processing conditions on the microstructure, porosity, and mechanical properties of Ti-6Al-4V repair fabricated by directed energy deposition. *J. Mater. Process. Technol.* **2019**, *264*, 172–181. [CrossRef]
83. Wolff, S.J. *Laser-Matter Interactions in Directed Energy Deposition*; Northwestern University Press: Evanston, IL, USA, 2018.
84. Spalding, I.J. Applied laser tooling. *J. Mod. Opt.* **2007**, *35*, 754–755. [CrossRef]
85. Wolff, S.J.; Lin, S.; Faierson, E.J.; Liu, W.K.; Wagner, G.J.; Cao, J. A framework to link localized cooling and properties of directed energy deposition (DED)-processed Ti-6Al-4V. *Acta Mater.* **2017**, *132*, 106–117. [CrossRef]
86. Ahsan, M.N.; Bradley, R.; Pinkerton, A.J. Microcomputed tomography analysis of intralayer porosity generation in laser direct metal deposition and its causes. *J. Laser Appl.* **2011**, *23*, 022009. [CrossRef]
87. Cunningham, R.; Nicolas, A.; Madsen, J.; Fodran, E.; Anagnostou, E.; Sangid, M.D.; Rollett, A. Analyzing the effects of powder and post-processing on porosity and properties of electron beam melted Ti-6Al-4V. *Mater. Res. Lett.* **2017**, *5*, 516–525. [CrossRef]
88. Rabin, B.H.; Smolik, G.R.; Korth, G.E. Characterization of entrapped gases in rapidly solidified powders. *Mater. Sci. Eng. A* **1990**, *124*, 1–7. [CrossRef]
89. Wang, Z.; Denlinger, E.; Michaleris, P.; Stoica, A.D.; Ma, D.; Beese, A.M. Residual stress mapping in Inconel 625 fabricated through additive manufacturing: Method for neutron diffraction measurements to validate thermomechanical model predictions. *Mater. Des.* **2017**, *113*, 169–177. [CrossRef]
90. Rafi, H.K.; Pal, D.; Patil, N.; Starr, T.L.; Stucker, B.E. Microstructure and Mechanical Behavior of 17-4 Precipitation Hardenable Steel Processed by Selective Laser Melting. *J. Mater. Eng. Perform.* **2014**, *23*, 4421–4428. [CrossRef]
91. Wang, Z.; Beese, A.M. Effect of chemistry on martensitic phase transformation kinetics and resulting properties of additively manufactured stainless steel. *Acta Mater.* **2017**, *131*, 410–422. [CrossRef]
92. Mukherjee, T.; Zuback, J.S.; De, A.; DebRoy, T. Printability of alloys for additive manufacturing. *Sci. Rep.* **2016**, *6*, 1–8. [CrossRef] [PubMed]
93. Sandgren, H.R.; Zhai, Y.; Lados, D.A.; Shade, P.A.; Schuren, J.C.; Groeber, M.A.; Kenesei, P.; Gavras, A.G. Characterization of fatigue crack growth behavior in LENS fabricated Ti-6Al-4V using high-energy synchrotron x-ray microtomography. *Addit. Manuf.* **2016**, *12*, 132–141. [CrossRef]
94. Beese, A.M.; Carroll, B.E. Review of mechanical properties of Ti-6Al-4V made by laser based additive manufacturing using powder feedstock. *JOM* **2016**, *68*, 724–734. [CrossRef]
95. Carroll, B.E.; Palmer, T.A.; Beese, A.M. Anisotropic tensile behavior of Ti-6Al-4V components fabricated with directed energy deposition additive manufacturing. *Acta Mater.* **2015**, *87*, 309–320. [CrossRef]
96. Shamsaei, N.; Yadollahi, A.; Bian, L.; Thompson, S.M. An overview of direct laser deposition for additive manufacturing; Part II: Mechanical behavior, process parameter optimization and control. *Addit. Manuf.* **2015**, *8*, 12–35. [CrossRef]

97. Zuback, J.S.; DebRoy, T. The Hardness of Additively Manufactured Alloys. *Materials* **2018**, *11*, 2070. [CrossRef] [PubMed]
98. Arcella, F.G.; Froes, F.H. Producing titanium aerospace components from powder using laser forming. *JOM* **2000**, *52*, 28–30. [CrossRef]
99. Kobryn, P.A.; Semiatin, S.L. *Mechanical Properties of Laser-Deposited Ti-6Al-4V*; Air Force Research Laboratory: Hanscom, MA, USA, 2013.
100. Zhai, Y.; Galarraga, H.; Lados, D.A. Microstructure evolution, tensile properties, and fatigue damage mechanisms in Ti-6Al-4V alloys fabricated by two additive manufacturing techniques. *Procedia Eng.* **2015**, *114*, 658–666. [CrossRef]
101. Bian, L.; Thompson, S.M.; Shamsaei, N. Mechanical properties and microstructural features of direct laser-deposited Ti-6Al-4V. *JOM* **2015**, *67*, 629–638. [CrossRef]
102. Nalla, R.K.; Ritchie, R.O.; Boyce, B.L.; Campbell, J.P.; Peters, J.O. Influence of microstructure on high-cycle fatigue of Ti-6Al-4V: Bimodal vs. lamellar structures. *Metall. Mater. Trans. A* **2002**, *33*, 899–918. [CrossRef]
103. Prabhu, A.W.; Vincent, T.; Chaudhary, A.; Zhang, W.; Babu, S.S. Effect of microstructure and defects on fatigue behaviour of directed energy deposited Ti–6Al–4V. *Sci. Technol. Weld. Join.* **2015**, *20*, 659–669. [CrossRef]
104. Kobryn, P.A.; Semiatin, S. Mechanical properties of laser-deposited Ti-6Al-4V. In *Solid Freeform Fabrication Proceedings*; Landes Bioscience: Austin, TX, USA, 2001.
105. Zhai, Y.; Lados, D.A.; Brown, E.J.; Vigilante, G.N. Fatigue crack growth behavior and microstructural mechanisms in Ti-6Al-4V manufactured by laser engineered net shaping. *Int. J. Fatigue* **2016**, *93*, 51–63. [CrossRef]
106. Li, C.; Liu, Z.Y.; Fang, X.Y.; Guo, Y.B. Residual stress in metal additive manufacturing. *Procedia CIRP* **2018**, *71*, 348–353. [CrossRef]
107. NIST. *Measurement Science for Additive Manufacturing Program*; NIST: Gaithersburg, MD, USA, 2013.
108. Kenel, C.; Grolimund, D.; Li, X.; Panepucci, E.; Samson, V.A.; Sanchez, D.F.; Marone, F.; Leinenbach, C. In situ investigation of phase transformations in Ti-6Al-4V under additive manufacturing conditions combining laser melting and high-speed micro-X-ray diffraction. *Sci. Rep.* **2017**, *7*, 1–10. [CrossRef]
109. Wolff, S.J.; Wu, H.; Parab, N.; Zhao, C.; Ehmann, K.F.; Sun, T.; Cao, J. In-situ high-speed X-ray imaging of piezo-driven directed energy deposition additive manufacturing. *Sci. Rep.* **2019**, *9*, 1–14. [CrossRef]
110. Koester, L.W.; Taheri, H.; Bigelow, T.A.; Bond, L.J.; Faierson, E.J. In-situ acoustic signature monitoring in additive manufacturing processes. *AIP Conf. Proc.* **2018**, *1949*, 020006. [CrossRef]
111. Wang, F.; Mao, H.; Zhang, D.; Zhao, X.; Shen, Y. Online study of cracks during laser cladding process based on acoustic emission technique and finite element analysis. *Appl. Surf. Sci.* **2008**, *255*, 3267–3275. [CrossRef]
112. Griffith, M.L.; Schlienger, M.E.; Harwell, L.D.; Oliver, M.S.; Baldwin, M.D.; Ensz, M.T.; Essien, M.; Brooks, J.; Robino, C.V.; Smugeresky, J.E.; et al. Understanding thermal behavior in the LENS process. *Mater. Des.* **1999**, *20*, 107–113. [CrossRef]
113. Nassar, A.R.; Keist, J.S.; Reutzel, E.W.; Spurgeon, T.J. Intra-layer closed-loop control of build plan during directed energy additive manufacturing of Ti–6Al–4V. *Addit. Manuf.* **2015**, *6*, 39–52. [CrossRef]
114. Wang, L.; Felicelli, S.D.; Craig, J.E. Thermal modeling and experimental validation in the LENS™ process. In Proceedings of the 18th Solid Freeform Fabrication Symposium, Austin, TX, USA, 2007; pp. 100–111.
115. Hua, T.; Jing, C.; Xin, L.; Fengying, Z.; Weidong, H. Research on molten pool temperature in the process of laser rapid forming. *J. Mater. Process. Technol.* **2008**, *198*, 454–462. [CrossRef]
116. Tellez, A.G.M. *Fibre Laser Metal Deposition with Wire: Parameters Study and Temperature Control*; University of Nottingham: Nottingham, UK, 2010.
117. Yu, J.; Lin, X.; Wang, J.; Chen, J.; Huang, W. Mechanics and energy analysis on molten pool spreading during laser solid forming. *Appl. Surf. Sci.* **2010**, *256*, 4612–4620. [CrossRef]
118. Hu, D.; Kovacevic, R. Sensing, modeling and control for laser based additive manufacturing. *Int. J. Mach. Tools Manuf.* **2003**, *43*, 51–60. [CrossRef]
119. Gegel, M.L.; Bristow, D.A.; Landers, R.G. A Quadratic-optimal repetitive process controller for laser metal deposition. In Proceedings of the Annual American Control Conference (ACC), Seattle, WA, USA, 27–29 June 2018.
120. Thomas, M.; Baxter, G.J.; Todd, I. Normalised model based processing diagrams for additive layer manufacture of engineering alloys. *Acta Mater.* **2016**, *108*, 26–35. [CrossRef]

121. Kuriya, T.; Ryo, K.; Oda, Y.; Yasuhiro, K. Evaluation and Analysis of Generated Void in Directed Energy Deposition of Inconel 718. *J. Jpn. Soc. Precis. Eng.* **2018**, *84*, 371–377. [CrossRef]
122. Ferguson, J.B.; Schultz, B.F.; Moghadam, A.D.; Rohatgi, P.K. Semi-empirical model of deposit size and porosity in 420 stainless steel and 4140 steel using laser engineered net shaping. *J. Manuf. Process.* **2015**, *19*, 163–170. [CrossRef]
123. Sciammarella, F.; Salehi Najafabadi, B. Processing Parameter DOE for 316L Using Directed Energy Deposition. *J. Manuf. Mater. Process.* **2018**, *2*, 61. [CrossRef]
124. Mahamood, R.M.; Akinlabi, E.T. Processing Parameters Optimization for Material Deposition Efficiency in Laser Metal Deposited Titanium Alloy. *Lasers Manuf. Mater. Process.* **2016**, *3*, 9–21. [CrossRef]
125. Ahsan, M.R.; Kim, Y.R.; Ashiri, R.; Cho, Y.J.; Jeong, C.; Park, Y.D. Cold metal transfer (CMT) gmaw of zinc-coated steel. *Weld. J.* **2016**, *95*, 120–132.
126. Fathi, A.; Toyserkani, E.; Khajepour, A.; Durali, M. Prediction of melt pool depth and dilution in laser powder deposition. *J. Phys. D Appl. Phys.* **2006**, *39*, 2613–2623. [CrossRef]
127. Bax, B.; Rajput, R.; Kellet, R.; Reisacher, M. Systematic evaluation of process parameter maps for laser cladding and directed energy deposition. *Addit. Manuf.* **2018**, *21*, 487–494. [CrossRef]
128. Fan, Z.; Jambunathan, A.; Sparks, T.E.; Ruan, J.; Yang, Y.; Bao, Y.; Liou, F. Numerical simulation and prediction of dilution during laser deposition. In *Solid Freeform Fabrication Proceedings, Proceedings of Seventeenth Annual Solid Freeform Fabrication (SFF) Symposium, Austin, TX, USA, 14–16 August 2006*; University of Texas at Austin: Austin, TX, USA, 2006; pp. 532–545.
129. Kong, F.; Kovacevic, R. Modeling of heat transfer and fluid flow in the laser multilayered cladding process. *Metall. Mater. Trans. B Process Metall. Mater. Process. Sci.* **2010**, *41*, 1310–1320. [CrossRef]
130. Saqiba, S.; Urbanica, R.J.; Aggarwal, K. Analysis of laser cladding bead morphology for developing additive manufacturing travel paths. *Procedia CIRP* **2014**, *17*, 824–829. [CrossRef]
131. Baidridge, T.; Poling, G.; Foroozmehr, E.; Kovacevic, R.; Metz, T.; Kadekar, V.; Gupta, M.C. Laser cladding of Inconel 690 on Inconel 600 superalloy for corrosion protection in nuclear applications. *Opt. Lasers Eng.* **2013**, *51*, 180–184. [CrossRef]
132. Bhardwaj, T.; Shukla, M.; Paul, C.P.; Bindra, K.S. Direct energy deposition-laser additive manufacturing of titanium-molybdenum alloy: Parametric studies, microstructure and mechanical properties. *J. Alloy. Compd.* **2019**, *787*, 1238–1248. [CrossRef]
133. Ansari, M.; Mohamadizadeh, A.; Huang, Y.; Paserin, V.; Toyserkani, E. Laser directed energy deposition of water-atomized iron powder: Process optimization and microstructure of single-tracks. *Opt. Laser Technol.* **2019**, *112*, 485–493. [CrossRef]
134. Hofman, J.T.; De Lange, D.F.; Pathiraj, B.; Meijer, J. FEM modeling and experimental verification for dilution control in laser cladding. *J. Mater. Process. Technol.* **2011**, *211*, 187–196. [CrossRef]
135. Harooni, A.; Nasiri, A.M.; Gerlich, A.P.; Khajepour, A.; Khalifa, A.; King, J.M. Processing window development for laser cladding of zirconium on zirconium alloy. *J. Mater. Process. Technol.* **2016**, *230*, 263–271. [CrossRef]
136. Sun, Y.; Hao, M. Statistical analysis and optimization of process parameters in Ti6Al4V laser cladding using Nd:YAG laser. *Opt. Lasers Eng.* **2012**, *50*, 985–995. [CrossRef]
137. Li, R.; Li, Z.; Huang, J.; Zhu, Y. Dilution effect on the formation of amorphous phase in the laser cladded Ni-Fe-B-Si-Nb coatings after laser remelting process. *Appl. Surf. Sci.* **2012**, *258*, 7956–7961. [CrossRef]
138. Huang, Y.; Khamesee, M.B.; Toyserkani, E. A new physics based model for laser directed energy deposition (powder-fed additive manufacturing): From single-track to multi-track and multi-layer. *Opt. Laser Technol.* **2019**, *109*, 584–599. [CrossRef]
139. Kobryn, P.A.; Moore, E.H.; Semiatin, S.L. Effect of laser power and traverse speed on microstructure, porosity, and build height in laser-deposited Ti-6Al-4V. *Scr. Mater.* **2000**, *43*, 299–305. [CrossRef]
140. Ansari, M.; Shoja Razavi, R.; Barekat, M. An empirical-statistical model for coaxial laser cladding of NiCrAlY powder on Inconel 738 superalloy. *Opt. Laser Technol.* **2016**, *86*, 136–144. [CrossRef]
141. De Oliveira, U.; Ocelík, V.; De Hosson, J.T.M. Analysis of coaxial laser cladding processing conditions. *Surf. Coat. Technol.* **2005**, *197*, 127–136. [CrossRef]
142. Koike, M.; Martinez, K.; Guo, L.; Chahine, G.; Kovacevic, R.; Okabe, T. Evaluation of titanium alloy fabricated using electron beam melting system for dental applications. *J. Mater. Process. Technol.* **2011**, *211*, 1400–1408. [CrossRef]

143. Cui, Z.; Yang, B.; Li, R.K. Application of biomaterials in cardiac repair and regeneration. *Engineering* **2016**, *2*, 141–148. [CrossRef]
144. Xue, W.; Krishna, B.V.; Bandyopadhyay, A.; Bose, S. Processing and biocompatibility evaluation of laser processed porous titanium. *Acta Biomater.* **2007**, *3*, 1007–1018. [CrossRef] [PubMed]
145. Hofmann, D.C.; Roberts, S.; Otis, R.; Kolodziejska, J.; Dillon, R.P.; Suh, J.O.; Shapiro, A.A.; Liu, Z.K.; Borgonia, J.P. Developing gradient metal alloys through radial deposition additive manufacturing. *Sci. Rep.* **2014**, *4*, 5357. [CrossRef] [PubMed]
146. Korinko, P.; Adams, T. Laser engineered net shaping for repair and hydrogen compatibility. *Weld. J.* **2011**, *90*, 171–181.
147. Chaudhari, R.; Ingle, A.; Kalita, K. Stress Analysis of Dissimilar Metal Weld between Carbon Steel and Stainless Steel formed by Transition Grading Technique. *Mater. Today Proc.* **2015**, *2*, 1657–1664. [CrossRef]
148. Brentrup, G.; Leister, B.; Snowden, B.; DuPont, J.; Grenestedt, J. Preventing dissimilar metal weld failures: Application of new functionally graded transition joints. *Proc. Mater. Sci. Technol.* **2009**, *2009*, 2554–2562.
149. Sexton, L.; Lavin, S.; Byrne, G.; Kennedy, A. Laser cladding of aerospace materials. *J. Mater. Process. Technol.* **2002**, *122*, 63–68. [CrossRef]
150. Zhou, S.; Zeng, X.; Hu, Q.; Huang, Y. Analysis of crack behavior for Ni based WC composite coatings by laser cladding and crack-free realization. *Appl. Surf. Sci.* **2008**, *255*, 1646–1653. [CrossRef]
151. Mackwood, A.P.; Crafer, R.C. Thermal modelling of laser welding and related processes: A literature review. *Opt. Laser Technol.* **2005**, *37*, 99–115. [CrossRef]
152. Li, L. Advances and characteristics of high-power diode laser materials processing. *Opt. Lasers Eng.* **2000**, *34*, 231–253. [CrossRef]
153. Das, M.; Bhattacharya, K.; Dittrick, S.A.; Mandal, C.; Balla, V.K.; Sampath Kumar, T.S.; Bandyopadhyay, A.; Manna, I. In situ synthesized TiB-TiN reinforced Ti6Al4V alloy composite coatings: Microstructure, tribological and in-vitro biocompatibility. *J. Mech. Behav. Biomed. Mater.* **2014**, *29*, 259–271. [CrossRef] [PubMed]
154. Han, L.; Phatak, K.M.; Liou, F.W. Modeling of laser cladding with powder injection. *Metall. Mater. Trans. B* **2007**, *35*, 1139–1150. [CrossRef]
155. Paul, C.P.; Jain, A.; Ganesh, P.; Negi, J.; Nath, A.K. Laser rapid manufacturing of Colmonoy-6 components. *Opt. Lasers Eng.* **2006**, *44*, 1096–1109. [CrossRef]
156. Calleja, A.; Tabernero, I.; Fernández, A.; Celaya, A.; Lamikiz, A.; López De Lacalle, L.N. Improvement of strategies and parameters for multi-axis laser cladding operations. *Opt. Lasers Eng.* **2014**, *56*, 113–120. [CrossRef]
157. Zhang, J.; Liou, F. Adaptive Slicing for a Multi-Axis Laser Aided Manufacturing Process. *J. Mech. Des.* **2004**, *126*, 254–261. [CrossRef]
158. Liu, Z.; Cong, W.; Kim, H.; Ning, F.; Jiang, Q.; Li, T.; Zhang, H.C.; Zhou, Y. Feasibility exploration of superalloys for AISI 4140 steel repairing using laser engineered net shaping. *Procedia Manuf.* **2017**, *10*, 912–922. [CrossRef]
159. Bi, G.; Gasser, A. Restoration of nickel-base turbine blade knife-edges with controlled laser aided additive manufacturing. *Phys. Procedia* **2011**, *12*, 402–409. [CrossRef]
160. Díaz, E.; Amado, J.M.; Montero, J.; Tobar, M.J.; Yáñez, A. Comparative study of Co based alloys in repairing low Cr-Mo steel components by laser cladding. *Phys. Procedia* **2012**, *39*, 368–375. [CrossRef]
161. Raju, R.; Duraiselvam, M.; Petley, V.; Verma, S.; Rajendran, R. Microstructural and mechanical characterization of Ti6Al4V refurbished parts obtained by laser metal deposition. *Mater. Sci. Eng. A* **2015**, *643*, 64–71. [CrossRef]
162. Knoll, H.; Ocylok, S.; Weisheit, A.; Springer, H.; Jägle, E.; Raabe, D. Combinatorial Alloy Design by Laser Additive Manufacturing. *Steel Res. Int.* **2017**, *88*, 1–11. [CrossRef]
163. Gwalani, B.; Soni, V.; Waseem, O.A.; Mantri, S.A.; Banerjee, R. Laser additive manufacturing of compositionally graded AlCrFeMoV*x* (*x* = 0 to 1) high-entropy alloy system. *Opt. Laser Technol.* **2019**, *113*, 330–337. [CrossRef]
164. Buchanan, C.; Gardner, L. Metal 3D printing in construction: A review of methods, research, applications, opportunities and challenges. *Eng. Struct.* **2019**, *180*, 332–348. [CrossRef]
165. Cortina, M.; Arrizubieta, J.I.; Ruiz, J.E.; Ukar, E.; Lamikiz, A. Latest developments in industrial hybrid machine tools that combine additive and subtractive operations. *Materials* **2018**, *11*, 2583. [CrossRef] [PubMed]

166. Hönnige, J.R.; Colegrove, P.A.; Ahmad, B.; Fitzpatrick, M.E.; Ganguly, S.; Lee, T.L.; Williams, S.W. Residual stress and texture control in Ti-6Al-4V wire + arc additively manufactured intersections by stress relief and rolling. *Mater. Des.* **2018**, *150*, 193–205. [CrossRef]
167. Ning, F.; Hu, Y.; Liu, Z.; Cong, W.; Li, Y.; Wang, X. Ultrasonic Vibration-Assisted Laser Engineered Net Shaping of Inconel 718 Parts: A Feasibility Study. *Procedia Manuf.* **2017**, *10*, 771–778. [CrossRef]

MDPI
St. Alban-Anlage 66
4052 Basel
Switzerland
Tel. +41 61 683 77 34
Fax +41 61 302 89 18
www.mdpi.com

MDPI Books Editorial Office
E-mail: books@mdpi.com
www.mdpi.com/books

www.ingramcontent.com/pod-product-compliance
Ingram Content Group UK Ltd.
Pitfield, Milton Keynes, MK11 3LW, UK
UKHW062009041026
15591UKWH00017B/1245

9 783039 284481